AF389013

Biofiltration and Physicochemical Filtration for Water Treatment

Biofiltration and Physicochemical Filtration for Water Treatment

Editor

Francisco Osorio

MDPI • Basel • Beijing • Wuhan • Barcelona • Belgrade • Manchester • Tokyo • Cluj • Tianjin

Editor
Francisco Osorio
University of Granada
Spain

Editorial Office
MDPI
St. Alban-Anlage 66
4052 Basel, Switzerland

This is a reprint of articles from the Special Issue published online in the open access journal *Water* (ISSN 2073-4441) (available at: https://www.mdpi.com/journal/water/special_issues/ biofiltration).

For citation purposes, cite each article independently as indicated on the article page online and as indicated below:

LastName, A.A.; LastName, B.B.; LastName, C.C. Article Title. *Journal Name* **Year**, *Volume Number*, Page Range.

ISBN 978-3-0365-1445-1 (Hbk)
ISBN 978-3-0365-1446-8 (PDF)

© 2021 by the authors. Articles in this book are Open Access and distributed under the Creative Commons Attribution (CC BY) license, which allows users to download, copy and build upon published articles, as long as the author and publisher are properly credited, which ensures maximum dissemination and a wider impact of our publications.

The book as a whole is distributed by MDPI under the terms and conditions of the Creative Commons license CC BY-NC-ND.

Contents

About the Editor

Francisco Osorio (Prof. Dr.) has been Full Professor at the School of Civil Engineers, University of Granada, Spain from 2016. He was Associate Professor at the same university from 2001. Francisco Osorio's Biofiltration Research Group dates back to the 1990s, when they initiated a line of research on the study of this technology for the treatment of urban wastewater. Subsequently, they have developed several doctoral theses and published more than 80 articles in specialized journals of great prestige worldwide (indexed in international databases), on different applications of biofiltration and physicochemical filtration, for the treatment of diverse effluents, in the field of drinking water and wastewater, both urban and industrial. In all these works, the pilot plant studies have been complemented with the analysis of the biomass and the space–time monitoring of their biodiversity, as well as with the characterization of the most relevant species of the functional microbial groups of participating bacteria in these processes, through the development of, among others, molecular biology techniques such as PCR, qPCR or massive DNA sequencing. Prof. Francisco Osorio has advised different multinational companies on biofiltration processes in real full-scale plants; he has also developed numerous research projects on the subject. He is the inventor of a patent on nitrogen autotrophic removal by means of biofiltration, the first worldwide on this process developed in a fixed bed biofilter.

Preface to "Biofiltration and Physicochemical Filtration for Water Treatment"

Biofiltration is a technology of great interest since the costs of installation and, above all, exploitation costs are much lower than those associated with other technologies based on physical–chemical processes. There are certain applications where the experience with biofiltration has been more widespread, but there are others where the application of biological systems has been traditionally minor. Nowadays, the use of biofiltration is increasing every day.

On the other hand, the physicochemical filtration process is a successful technology in numerous applications in the field of water treatment. In many cases, the mechanisms for removing contaminants that develop in a filter are complex. Thus, its efficiency may be due to the simple physicochemical mechanisms that characterize any filtration process, but the development of biofilms on the surface of the filtering materials is also common, so, in part, the efficiency can be achieved biologically. In this sense, the discussion about the mechanisms for removing contaminants in each case is significantly interesting.

This issue of the journal is focused on the treatment of different types of effluents through filtration: Drinking water and wastewater. Different technologies are analysed: Filtration through biochar from agricultural by-products; biological active carbon (BAC); electroadsorption using a commercial granular activated carbon as the adsorbent; filtration through sand, anthracite and expanded clay; granular activated carbon (GAC) as part of a tertiary treatment for wastewater reuse.

Francisco Osorio
Editor

Article

Biochar from Agricultural by-Products for the Removal of Lead and Cadmium from Drinking Water

Edgar Pineda Puglla [1], Diana Guaya [2], Cristhian Tituana [3], Francisco Osorio [4] and María J. García-Ruiz [4,*]

[1] Department of Geology, Mines and Civil Engineering, UTPL, Universidad Técnica Particular de Loja, Barrio San Cayetano Alto, Marcelino Champagnat Street, s/n, Loja 110107, Ecuador; eipineda@utpl.edu.ec
[2] Department of Chemical, UTPL, Universidad Técnica Particular de Loja, Barrio San Cayetano Alto, Marcelino Champagnat Street, s/n, Loja 110107, Ecuador; deguaya@utpl.edu.ec
[3] Civil Engineering Degree, UTPL, Universidad Técnica Particular de Loja, Barrio San Cayetano Alto, Marcelino Champagnat Street, s/n, Loja 110107, Ecuador; catituana1@utpl.edu.ec
[4] Department of Civil Engineering, Institute of Water, University of Granada, Ramón y Cajal Street, 4, 18071 Granada, Spain; fosorio@ugr.es
* Correspondence: mjgruiz@ugr.es; Tel.: +34-9-5824-9463

Received: 29 August 2020; Accepted: 14 October 2020; Published: 20 October 2020

Abstract: This study reports the adsorption capacity of lead Pb^{2+} and cadmium Cd^{2+} of biochar obtained from: peanut shell (BCM), "chonta" pulp (BCH) and corn cob (BZM) calcined at 500, 600 and 700 °C, respectively. The optimal adsorbent dose, pH, maximum adsorption capacity and adsorption kinetics were evaluated. The biochar with the highest Pb^{2+} and Cd^{2+} removal capacity is obtained from the peanut shell (BCM) calcined at 565 °C in 45 min. The optimal experimental conditions were: 14 g L^{-1} (dose of sorbent) and pH between 5 and 7. The sorption experimental data were best fitted to the Freundlich isotherm model. High removal rates were obtained: 95.96% for Pb^{2+} and 99.05. for Cd^{2+}. The BCH and BZM revealed lower efficiency of Pb^{2+} and Cd^{2+} removal than BCM biochar. The results suggest that biochar may be useful for the removal of heavy metals (Pb^{2+} and Cd^{2+}) from drinking water.

Keywords: peanut shell; "chonta" pulp; corn cob; biochar; adsorption; lead; cadmium; drinking water

1. Introduction

Water pollution by wastewater discharges into rivers or bodies of water by anthropogenic activities has increased due to population growth [1,2]. Heavy metals in water promote toxicity, and they are not biodegradable [3,4]. Low concentration of heavy metals has a great impact on human health and aquatic life. They can cause respiratory problems, weakening of the immune system, damage to the kidneys or liver, genetic and neurological alterations and death [5]. Lead (Pb^{2+}) and cadmium (Cd^{2+}) are abundant in nature; however, they are very toxic. Pb^{2+} and Cd^{2+} are incorporated into the food chain in low concentrations by water systems, affecting wildlife and people [6].

In South America, some rivers that supply drinking water to cities contain Pb^{2+} in high concentrations. The Rímac River in Lima, Peru, in 2009, registered a concentration of 2.15 mg L^{-1} Pb^{2+} [7]. In 2017, the Rímac river maintained a high Pb^{2+} concentration (2.064 mg L^{-1}) and also reported a Cd^{2+} concentration of 0.038 mg L^{-1} [8]. In Ecuador, some rivers contain heavy metals from mining [9]. The Puyango river, located between Loja and El Oro provinces at southern Ecuador, reported an average content of 0.77 mg L^{-1} of Pb^{2+}. However, the water from Puyango river is used for agricultural application and human consumption by northern Peru.

There are some technologies for heavy metal removal from water, such as chemical precipitation, ion exchange, electrochemical treatment, membrane technologies, reverse osmosis, electrodialysis, nanofiltration, coagulation and adsorption [10–15]. Adsorption is an efficient, simple and low cost technology for heavy metal removal from water [16,17]. A low cost, renewable and high heavy metal sorption capacity characterizes an efficient sorbent [18]. The activated carbon is efficient for heavy metal adsorption; however, it is expensive and unprofitable for water treatment. So, the use of agricultural by-products as a bioadsorbent-type biochar has become attractive for some researchers [19,20].

Biochar from vegetal sources is used for carbon sequestration due to its stability and physicochemical properties [21]. The biochar properties depend mainly on the source raw material and the calcination temperature. Biochar contains mineral and carbon fractions. The formation of aromatic structures and oxygenated functional groups (Oxygen Functional Groups, OFG) are controlled by the calcination temperature [22]. Biochar is also an effective sorbent for nutrient removal from water due to the high specific surface, porous structure and high cation and anion exchange capacity [23].

The aim of this study is to use the agricultural by-products: peanut shell, "chonta" pulp and corn cob, to transform them into biochar for Pb^{2+} and Cd^{2+} removal from water. The peanut shell, "chonta" pulp and corn cob were selected because they are agricultural residues that are found in abundance in the location. Not much information has been found about previous studies using "chonta" pulp for biochar obtention and adsorption of heavy metals. The objectives of the present study are: (i) determine the optimal calcination conditions of the agricultural by-products for biochar obtention and (ii) evaluate the efficiency of the biochar for Pb^{2+} and Cd^{2+} removal by batch experiments.

2. Materials and Methods

2.1. Raw Materials Collection

The agricultural by-products: peanut shells, "chonta" pulp and corn cob. The "chonta" (Bactris gasipaes) belongs to the Araceae family. The "chonta" grows in the tropical and subtropical regions of the American continent [24,25]. "Chonta" is fruit which is part of the nutritional diet of the indigenous tribes from the Amazon. The peanut (Arachis hypogaea) is a legume of the Fabaceae family which is originally from South America [26]. The corn (Zea Mays) belongs to the Graminea family [27]. The raw material was acquired in the southern region of Ecuador. The peanut shell and the corn cob were collected at Loja province (3°52′23″ S–79°38′27″ W). The "chonta" fruit was obtained from Ecuadorian Amazon at Zamora Chinchipe province (4°4.011′ S–78°57.293′ W).

2.2. Calcination of Biomass

The "chonta" pulp was separated from the walnut using a 1.5 kg mortar. The peanut shell and the corn cob were processed in situ from the harvesting tasks. The raw materials were washed with deionized water to remove impurities. After, raw materials were dried at 105 °C for 24 h [28,29].

The optimal calcination temperature and time were determined with a thermogravimetric analysis. A mass (10 g) of each biomass was calcined from 30 °C to 800 °C at 10 °C min^{-1} rate [30–32]. The carbon content and yield were evaluated for the obtained biochar. The biomass was calcined until 500 °C maintaining it at a constant during the 30 min [33–35]. Furthermore, the calcination was evaluated during 45 and 60 min until 600 and 700 °C, respectively [36].

After calcination, biochar was triturated pass through a mechanical sieve equipment (Retsch AS200). There was obtained biochar sizes in the range of 4.76 mm to 2 mm, sieve ASTM No. 4 and ASTM No. 10, respectively. The biochar particles were washed with deionized distilled water to remove impurities and ash [29,37], finally biochar were dried at 105 °C during 60 min [28,38].

2.3. Biochar Characterization

The moisture content, volatile matter, ash and fixed carbon were determined, according to the ASTM D3174-12 (Standard Test Method for Ash in the Analysis Sample of Coal and Coke from Coal) standard [34,39]. The yield was determined by weight mass difference between the biomass and the biochar obtained after calcination [40]. The moisture content was determined by drying 1 g of biochar at 105 °C for 180 min, then the sample was placed in a desiccator until the final weight was registered [41].

The morphology and composition of the biochar were studied in a Scanning Electron Microscopy coupled to the Energy Dispersive Spectroscopy system.

To biochar specific surface was determined by the nitrogen gas adsorption method. An automatic adsorption analyzer (Micrometrics) was used. Trials were performed with three replicates and average values are reported.

The ash content was determined using 0.10 g of dry biochar. The sample was introduced into a muffle preheated at 650 °C between 3 h and 16 h. The calcination was completed when constant weight was obtained [42]. The volatile material was determined by weighing 1 g of biochar, which is preheated in a muffle at 950 °C for 7 min. Finally, the biochar weight correspond to the non-volatile compounds [43]. The moisture, volatile material, ash and fixed carbon of biochar is equal to 100% of the carbon composition. The fixed carbon was determined by mass balance from 100% of the carbon composition, the percentage of moisture, ash and volatile material [39].

The weight: volume ratio as the apparent density was determined according to ASTM D2854 - 09 (Standard Test Method for Apparent Density of Activated Carbon). An electro vibrator was used with a uniform flow range of 0.75 cm^3 s^{-1} to 1 cm^3 s^{-1}, to improve the density results [44].

Finally, the use of a Hirox KH 8700 digital microscope and Labscope software, the pore size on the grain surface was measured.

2.4. Evaluation of Pb^{2+} and Cd^{2+} Removal from Aqueous Solution

A nomenclature was used to identify each biochar obtained from: peanut shell (BCM), "chonta" (BCH) and corn cob (BZM). The adsorption capacity was determined using Equation (1) [32].

$$q_e = \frac{(C_o - C_e)}{m} \times W \qquad (1)$$

where q_e (mg g^{-1}) is the adsorption capacity; C_o (mg L^{-1}) is the initial concentration; C_e. (mg L^{-1}) is the equilibrium concentration; W (L) is the volume of Pb^{2+} or Cd^{2+} aqueous solution and m (g) is the sorbent mass. The removal percentage (%Removal) were determined by Equation (2) [32].

$$\%\text{Removal} = \frac{(C_o - C_e)}{C_o} \times 100 \qquad (2)$$

2.4.1. Effect of pH

The adsorption capacity and removal percentage were evaluated at pH 3, 5, 7 and 9 at room temperature of 18 °C. Synthetic solutions containing 2 mg L^{-1} Pb^{2+} and 2 mg L^{-1} Cd^{2+} at were prepared from Pb (NO$_3$)$_2$ (purity 99.5%, MERCK) and using a standard of 1000 mg L^{-1} Cd^{2+}, SIGMA – ALDRICH, respectively. The pH of the solution was adjusted using NH4OH at 0.1 mol L-1 or HCl 0.1 M [45]. The amount of biochar and adsorbate was the same for each pH value under evaluation. The flask was stirred for 45 min at 140 rpm. The equilibrated solution were filtered on a 45 μm filter paper and HNO$_3$ was added (0.1 mol L^{-1}) to avoid the precipitation of metal ions [28]. An inductively coupled plasma optical emission spectrometer (Perkin Elmer OPTIMA 8000) was used for determining the metals' concentration. The adsorption capacity and removal percentage were determined.

2.4.2. Effect of Adsorbent Dose

The amount of biochar was evaluated using concentrations of 8, 10, 12, 14 and 16 g L^{-1} of adsorbent at room temperature at 18 °C. 100 mL of solution containing 2 mg L^{-1} Pb^{2+} and 2 mg L^{-1} Cd^{2+} at pH 5.

The flask was stirred for 45 min at 140 rpm. The equilibrated solution was filtered on a 45 μm filter paper. The equilibrium concentration C_e was determined.

2.4.3. Adsorption Isotherms

The initial concentrations of the synthetic Pb^{2+} and Cd^{2+} solutions were: 1, 2, 3 and 4 mg L^{-1}. 50 mL of synthetic solutions using 2 mg L^{-1} of sorbent were equilibrated during 45 min at 18 °C. The equilibrated solution was filtered on a 45 μm filter paper. The equilibrium concentration C_e was determined. The experimental data were fitted to de Langmuir and Freundlich isotherm models [28,44]. Langmuir model describe the chemical or monolayer adsorption and Freundlich will indicate whether it is a physical or multilayer adsorption. The Langmuir isotherm model is represented by Equation (3) [46].

$$q_e = \frac{q_{max} K_L C_e}{1 + K_L C_e} \tag{3}$$

where is maximum adsorption capacity required for the formation of monolayer; K_L (L mg^{-1}) is a Langmuir constant related to the affinity constant between the adsorbent and an adsorbate.

The Freundlich equation model is represented by Equation (4) [47].

$$q_e = K_F C_e^n \tag{4}$$

where K_F (L mg^{-1}) is the Freundlich adsorption constant, which characterizes the strength of adsorption; n (dimensionless) is a Freundlich intensity parameter.

2.4.4. Adsorption Kinetic

The adsorption kinetics of a 2 mg L^{-1} Pb^{2+} and 2 mg L^{-1} Cd^{2+} solution was performed at 5 °C and 18 °C. In total, 100 mL of a 2 mg L^{-1} Pb^{2+} and Cd^{2+} at 5 were equilibrated with biochar at 140 rpm. A 5 mL aliquot was taken from the solution during equilibration at 5, 15, 30, 45 and 60 min [37]. The equilibrated solution was filtered on a 45 μm filter paper. The equilibrium concentration C_e was determined. The experimental data were fitted to the pseudo-first order [48] and the pseudo-second order [49] kinetic model. The adsorption of metal ions as a function of time was determined by Equation (5) [37].

$$q_t = \frac{W(C_o - C_t)}{m} \tag{5}$$

where q_t (mg g^{-1}) is the sorption capacity at t time; C_t (mg L^{-1}) is the concentration of an adsorbate after a contact time t (min). The Lagergren pseudo-first order model is represented by Equation (6) [48].

$$\frac{dq}{dt} = K_1\left(q_e - q_t\right) \rightarrow q_t = q_e\left(1 + e^{-K_1 t}\right) \tag{6}$$

where K_1 (min^{-1}) is the first order rate constant; t (min) is the contact time. The Ho Model or pseudo-second order is denoted by Equation (7) [49].

$$\frac{dq}{dt} = K_2\left(q_e - q_t\right)^2 \tag{7}$$

Integrating the previous equation for the conditions of $q_t = 0$ for t $= 0$, Equation (8) [50].

$$q_t = \left(\frac{t}{\frac{1}{K_2 q_e^2} + \frac{t}{q_e}}\right) \tag{8}$$

where K_2 (g mg^{-1} min^{-1}) is the second order constant. The adsorption rate is denoted by Equation (9).

$$v = \frac{q_e}{t_e} \tag{9}$$

where v (mg g^{-1} min^{-1}) is the adsorption rate; t_e (min) is the equilibrium time determined by the kinetic.

3. Results

3.1. Calcination Conditions

The curves for the fixed carbon content at 30, 45 and 60 min are presented by Figure 1. They are identified as FC (30 min), FC (45 min) and FC (60 min). The yield is identified as Y (30 min), Y (45 min) and Y (60 min), respectively.

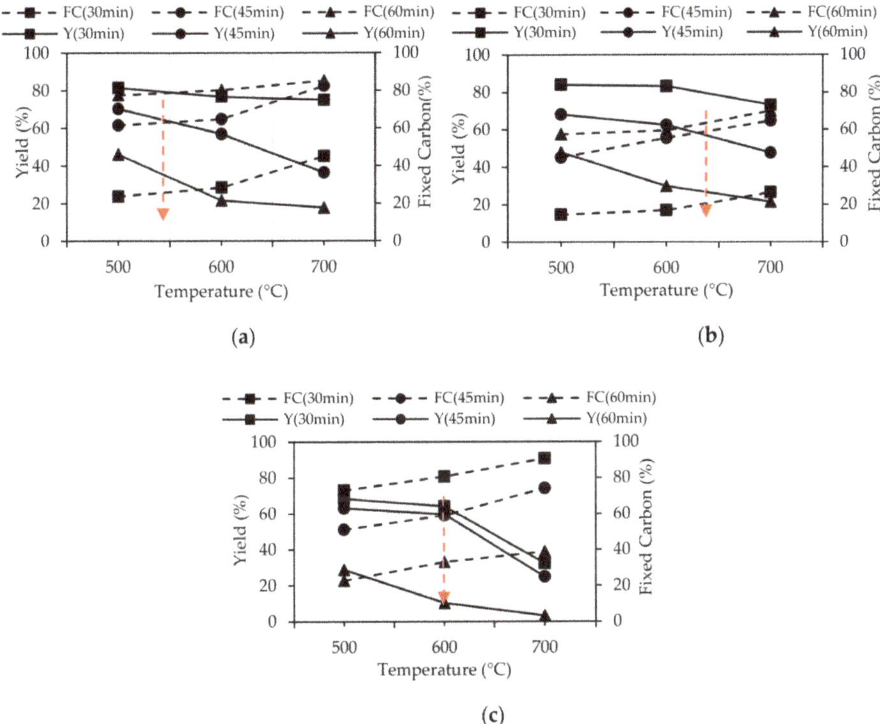

Figure 1. Yield vs. fixed carbon of biochar: (a) peanut shell (BCM), (b) "chonta" pulp (BCH) and (c) corn cob (BZM).

The intersection of Y and FC curves is the optimal carbonization point. For the BCM, the optimal calcination point was determined at 565 °C for 45 min. The BCH is optimally calcinated at 630 °C for 45 min, and the BZM is optimally calcined at the 600 ° C for 45 min. At 30 and 60 min, biochar present high content of FC and a low Y.

These results have been compared with other studies developed with biochar obtained from organic material. N'goran et al. [37] calcined cashew shells at 500 °C for 240 min and walnut shells at 450 °C for 120 min in an electric oven. Colpas et Al. [39] calcined the corn cob at 400 °C for 60 min in a multipurpose furnace. Castellar [51] reports calcination temperatures of 530 °C for 30 min for the cassava shell, carried out in a muffle furnace. Therefore, the carbonization temperature obtained for the organic materials used is in the order of the values reported in similar studies, considering it ideal for the transformation of these biomass.

3.2. Biochar Characterization

The result of biochar characterization is presented in Table 1. Peanut shell biochar obtained at 565 °C, "chonta" pulp biochar obtained at 630 °C and corn cob biochar obtained at 600 °C. The moisture content of biochar is according to the recommended values suggested according to the ASTM D2867-04, between 2%–15%.

Table 1. Physicochemical characteristics of biochar.

Parameter	Units	BCM	BCH	BZM
Apparent density	$g\,cm^{-3}$	0.12	0.17	0.11
Moisture content	%	6.81	5.98	5.23
Volatile material Mv	%	24.49	34.62	19.99
Ash content C_c	%	5.85	10.48	4.22
Fixed carbon Cf	%	62.85	48.92	70.22
Specific surface	$m^2\,g^{-1}$	1224	652.8	778.3
Pore size	μm	21.11	30.62	28.44
Effective size (D_{10})	mm	1.45	2.28	2.11
Yield Y	%	60.16	78.64	61.23

The apparent density of the biochar is in the order of the range established by the ASTM D2854-09 standard, from 0.26–0.65 g cm^{-3}. The apparent density of the biochar obtained is low, so the mass of these adsorbents should be less in a batch device. This fact can be favorable because fewer particles increase the porosity and a better contact between the adsorbent's surface and the adsorbate can occur. This avoids overlapping of adsorption sites due to large adsorbent masses [44]. The biochars obtained are in accordance with the AWWA B604-90 (Standard for granular activated carbon) [52]. This Standard recommends that the particle of the granular carbon for water treatment should have an effective size between 0.4 to 3.3 mm.

The ASTM D3175-20 (Standard Test Method for Volatile Matter in the Analysis Sample of Coal and Coke) considers Mv is optimal between 21.25% and 28.84%. ASTM D3175-20 (Standard Test Method for Volatile Matter in the Analysis Sample of Coal and Coke) establishes a range of 21.25%–28.84% for Mv. The Mv of BCH is higher than BCM and BZM. A biochar with a low content of volatile material is not very combustible; the lower the amount of Mv, the higher the fixed carbon content, an aspect that will favor at the time of adsorption [39,53,54].

The Cc of the biochar are in accordance with the ASTM D-2866-11 (Standard Test Method for Total Ash Content of Activated Carbon) standard establishes a range from 3% to 15%. A high ash content indicates that the biochar is fragile to carbonize but the low Cf content obtained affect its specific surface the adsorption capacity [55,56]. The higher specific surface the better the adsorption capacity. The specific surface of the BCM was higher and had a higher adsorption capacity than the other materials tested. The specific surface of the biochar was compared with other similar materials (Table 2).

Table 2. Specific surface area of various activated carbons and biochar.

Sorbent	$S\ (m^2\ g^{-1})$	Reference
Ripened Tea Leaf (MTL)	1313	[56]
Macore fruit	229.5	[57]
Palm oil mill effluent	59.19	[58]
Coconut shell	1135	[59]
Cashew shell	395.0	[60]
Cashew shell	984.0	[61]
Shea shell	768	[37]
Cashew shell	512	[37]
Peanut shell (BCM)	1224	This study
"Chonta" pulp (BCH)	652	This study
Corn cob (BZM)	778	This study

3.3. Sorbent Chemical Characteristics

The morphology and chemical compositions of the adsorbents were analyzed using a scanning electron microscope (SEM) coupled Energy Dispersive X-ray spectroscopy (EDX) (EDX) (Figure 2a–c). BZM contains mainly silicon, the whitish coloration can be leathery materials from the plant that generated the carbon. BCH contains a lot of calcium sulphate, gypsum, some phosphorus (typical of organic materials) and a little silicon.

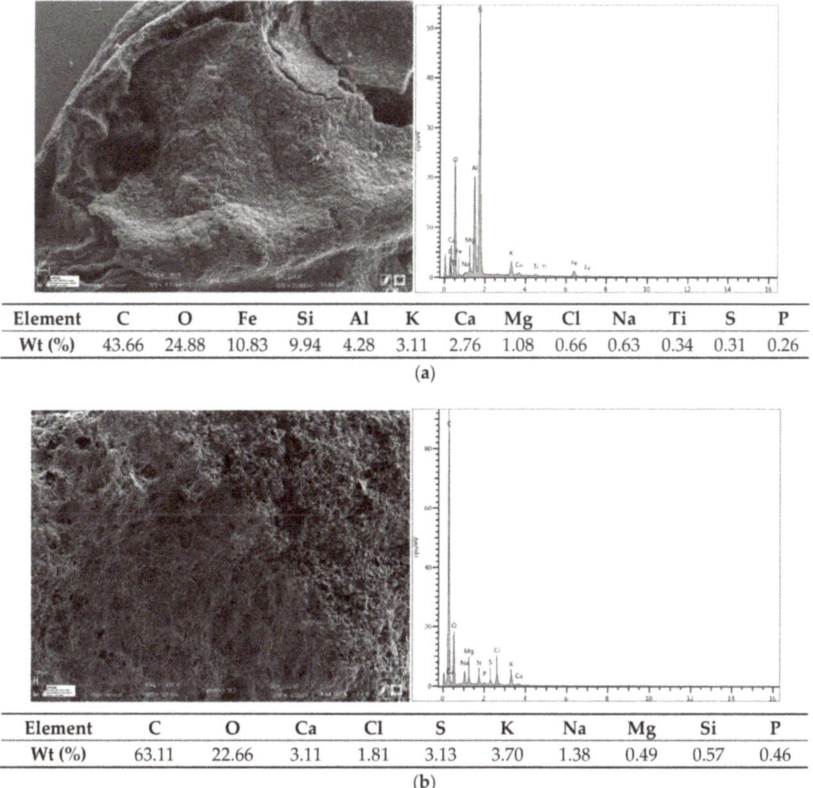

Element	C	O	Fe	Si	Al	K	Ca	Mg	Cl	Na	Ti	S	P
Wt (%)	43.66	24.88	10.83	9.94	4.28	3.11	2.76	1.08	0.66	0.63	0.34	0.31	0.26

(a)

Element	C	O	Ca	Cl	S	K	Na	Mg	Si	P
Wt (%)	63.11	22.66	3.11	1.81	3.13	3.70	1.38	0.49	0.57	0.46

(b)

Figure 2. *Cont.*

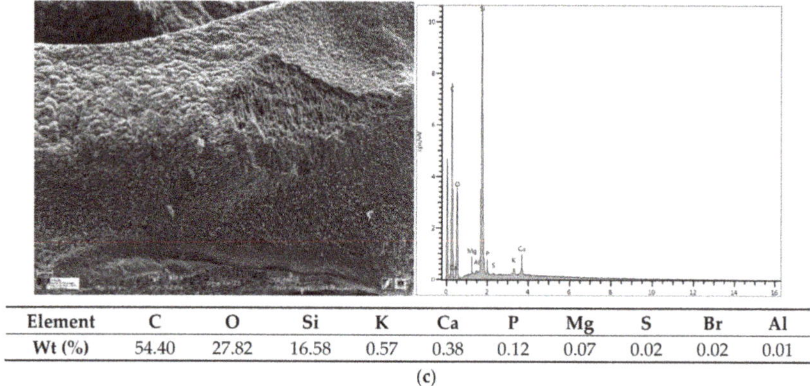

Element	C	O	Si	K	Ca	P	Mg	S	Br	Al
Wt (%)	54.40	27.82	16.58	0.57	0.38	0.12	0.07	0.02	0.02	0.01

(c)

Figure 2. SEM photographs and EDS spectra of adsorbents, (**a**) BCM ≈ 565 °C, (**b**) BCH ≈ 630 °C, (**c**) BZM ≈ 605 °C.

3.4. Adsorption as A Function of pH

The pH is determinant for the adsorption of heavy metals for biochar, since it shows that the adsorption occurs through electrostatic attraction. In Figure 3 is represented the adsorption as a function of pH for the three biochars. The Pb^{2+} adsorption increases with the increase of pH from 3 to 5. However, when pH increase from 7 to 9 the Pb^{2+} adsorption decrease. The Cd^{2+} removal percentage is the same for the pH range evaluated. The optimal pH for the removal of Pb^{2+} and Cd^{2+} is performed the best at pH 5.

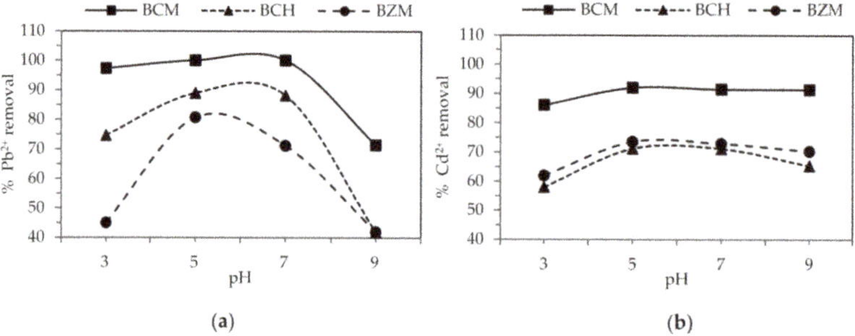

(a)   (b)

Figure 3. Percentage of (**a**) Pb^{2+} and (**b**) Cd^{2+} adsorption as a function of pH.

The adsorption decrease below pH 5 because an excess of hydrogen ions is generated, producing a competition with the positively charged metal ions towards the same places on the adsorbent surface [37]. Removal also decreases when pH is higher than 7, the pH increase promotes the formation of anionic hydroxide complexes that decrease the concentrations of free Pb^{2+} ions [62]. The occurrence of the hydrolysis reactions is represented by Equations (10)–(12):

$$[Pb^{2+} + OH^- \rightarrow Pb(OH)^+, -pk_a = 6.48] \tag{10}$$

$$[Pb(OH)^+ + OH^- \rightarrow Pb(OH)_2, -pk_a = 11.16] \tag{11}$$

$$[Pb(OH)_2 + OH^- \rightarrow Pb(OH)_{3-}, -pk_a = 14.16] \tag{12}$$

Kumar et al. [61] and Coelho et al. [63] determine pH 5 is the best for lead and cadmium removal using activated carbon obtained from cashew nuts in India and Brazil, respectively. The behavior of the biochar evaluated in this study represents an excellent possibility for drinking water treatment with typical pH values between 6.5 and 7.5.

3.5. Effect of Adsorbent Dose

The BCM using 14 g L^{-1} allowed 86% of Pb^{2+} removal, but BCH and BZM with 12 g L^{-1} removes a maximum of 74% and 85%, respectively (Figure 4). There are no differences between the BCM doses used for Cd^{2+} removal. The BCH and BZM using 14g L^{-1} obtained over 90% of Cd^{2+} removal. The percentage of removal increases with mass biochar increase. The greater the biochar mass the more available spaces for adsorption. However, when the equilibrium is reached, no matter how much biochar is used, the removal percentage does not increase. The aggregation or partial agglomeration of the adsorbent particles in higher concentration promotes this behavior [6,64,65].

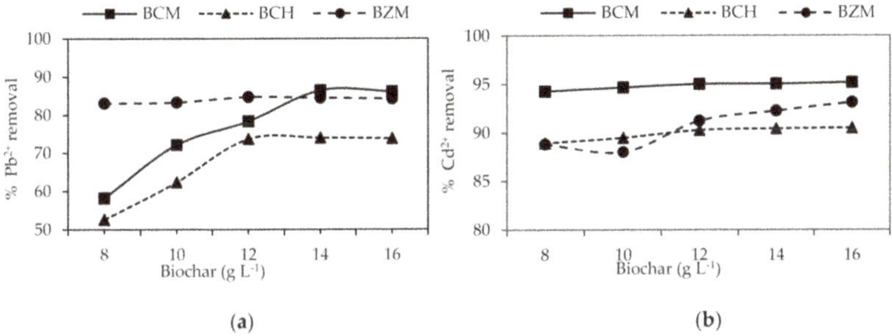

Figure 4. Removal percentage as a function of the biochar dose: (a) Pb^{2+} and (b) Cd^{2+}.

N'goran et al. [37] reported the optimum Pb^{2+} removal using activated carbon from cashews and shea nuts using 12 g L^{-1}. Coelho et al. [63] reported experimental essays using 12 g L^{-1} for activated carbon prepared from Brazilian cashew shell.

3.6. Adsorption Kinetics

The equilibrium sorption for Pb^{2+} and Cd^{2+} on the biochar was reached within 45 min at which was obtained the highest removal value (Figure 5).

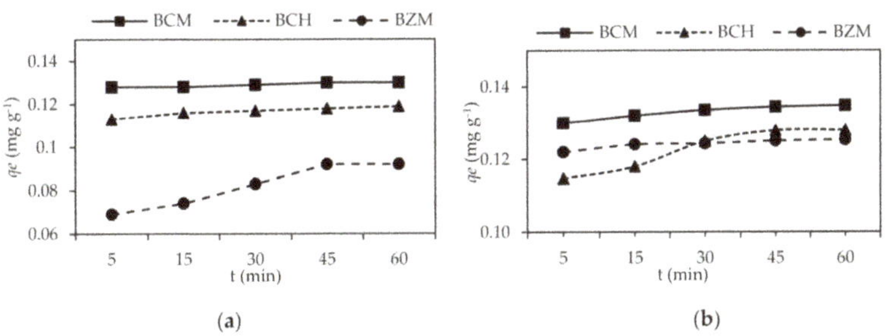

Figure 5. Biochar adsorption kinetics: (a) Pb^{2+} and (b) Cd^{2+}.

BCM developed the higher adsorption rate 0.0029 mg^{-1} g min^{-1} which developed the highest Pb^{2+} removal. The removal percentage of biochar: 95.96% for BCM, the 87.37% for BCH and 67.77% for BZM (Table 3). The adsorption capacity depends on the specific surface because the BCM (1224 m^2 g^{-1}) is higher than BZM and BCH. The BCM has a great surface to retain Pb^{2+} and Cd^{2+} from water.

Table 3. Comparison of the adsorption results obtained for the biochar.

Element	Biochar	C_o (mg L^{-1})	C_e (mg L^{-1})	Removal (%)	q_e (mg g^{-1})	Time (min)	Velocity (mg g^{-1} min^{-1})
Pb^{2+}	BCM	1.90	0.077	95.96	0.130	45	0.00289
	BCH	1.90	0.240	87.37	0.119	45	0.00263
	BZM	1.90	0.612	67.77	0.092	45	0.00204
Cd^{2+}	BCM	1.90	0.018	99.05	0.134	45	0.00299
	BCH	1.90	0.109	94.25	0.128	45	0.00284
	BZM	1.90	0.150	92.12	0.125	45	0.00278

The kinetic parameters of biochar evaluated in this study are summary in Table 4. So, the adsorption mechanism of Pb^{2+} and Cd^{2+} removal on biochar can be determined [65].

Table 4. Parameters of mathematical kinetic models.

Element	Biochar	First Order			Second Order		
		q_e (mg g^{-1})	K_1 (h^{-1})	R^2	q_e (mg g^{-1})	K_2 (g mg^{-1} h^{-1})	R^2
Pb^{2+}	BCM	0.019	6.094	0.658	0.130	2676.817	0.99995
	BCH	0.031	6.714	0.827	0.119	1334.948	0.99996
	BZM	0.099	8.199	0.836	0.096	187.224	0.99252
Cd^{2+}	BCM	0.028	6.247	0.754	0.142	1604.717	0.99998
	BCH	0.077	8.409	0.909	0.138	418.102	0.99999
	BZM	0.025	3.560	0.753	0.132	2879.693	0.99999

Taking into account the R^2 value near to 1 the experimental sorption data are best fitted to the pseudo-second order kinetic model. So, Pb^{2+} and Cd^{2+} removal on biochar is performed by chemisorption. Furthermore, Pb^{2+} and Cd^{2+} removal by biochar is governed by physisorption due to the electrostatic attraction previously discussed as an effect of pH [66].

Chemical sorption reactions occur through chemical bonds at specific functional groups which are irreversible. Previous studies attributed the Pb^{2+} and Cd^{2+} removal is performed by ion exchange reactions. The three biochar of this study contain some exchangeable metals on the surface such as: Na^+, K^+, Ca^{2+} and Mg^{2+} that allow the exchange. The adsorption of lead and cadmium by ion exchange reaction with those exchangeable ions can be described by Equation (13).

$$\text{B-Na}^+/\text{K}^+/\text{Ca}^{2+}/\text{Mg}^{2+} + \text{Pb}^{2+}/\text{Cd}^{2+} \rightarrow \text{B-Pb}^{2+}/\text{Cd}^{2+} + \text{Na+/K}^+/\text{Ca}^{2+}/\text{Mg}^{2+} \tag{13}$$

Furthermore, some complexation reactions occurred between Pb^{2+} and Cd^{2+} and the functional groups existing on the surfaces of coals. Biochar contains some organic groups containing oxygen that promote the Pb^{2+} and Cd^{2+} sorption on the biochar [67]. B-COOH and B-OH (B: biochar) represents the functional surface groups of BCM, BCH and BZM. The Pb^{2+} and Cd^{2+} adsorption is expected to occur by complexation reactions described in Equations (14) and (15) [29].

$$2\,\text{B-COOH} + \begin{matrix} \text{Pb}^{2+} \\ \text{Cd}^{2+} \end{matrix} \rightarrow \begin{matrix} 2\,\text{B-COOPb}^{2+} + 2\text{H}^+ \\ 2\,\text{B-COOCd}^{2+} + 2\text{H}^+ \end{matrix} \tag{14}$$

$$2\,\text{B-OH} + \begin{matrix} \text{Pb}^{2+} \\ \text{Cd}^{2+} \end{matrix} \rightarrow \begin{matrix} 2\,\text{B-OPb}^{2+} + 2\text{H}^+ \\ 2\,\text{B-OCd}^{2+} + 2\text{H}^+ \end{matrix} \tag{15}$$

Furthermore, it has been found in previous reports that some precipitation reactions may occur due to the components of biochar [28,68]. However, in this study we do not have evidence of this fact. The low Pb^{2+} and Cd^{2+} concentrations used for the study make it improbable to detect them by SEM-EDX characterization.

3.7. Adsorption Isotherms

The Pb^{2+} and Cd^{2+} removal by biochar increase with the increase of initial metal concentration. Taking into account the R^2, the experimental sorption data of both Pb^{2+} and Cd^{2+} on biochar are best fitted to Freundlich isotherm. The highest Pb^{2+} and Cd^{2+} sorption on biochar is obtained by BCM (Table 5).

Table 5. Fit to mathematical models of isotherms.

Element	Biochar	Langmuir			Freundlich		
		q_{MAX} (mg g^{-1})	K_L (L mg^{-1})	R^2	K_F (L mg^{-1})	n	R^2
	BCM	0.271	11.58	0.940	2.528	0.34	0.967
Pb^{2+}	BCH	0.245	4.34	0.763	0.532	0.76	0.999
	BZM	0.205	4.27	0.996	0.453	0.37	0.920
	BCM	1.038	1.67	0.189	0.816	0.86	0.839
Cd^{2+}	BCH	0.655	1.08	0.881	0.675	0.79	0.997
	BZM	0.857	0.37	0.534	0.0528	0.84	0.989

Accordingly, to Freundlich isotherm model the adsorption is performed by physical mechanisms (Figures 6 and 7). Physical adsorption denotes the existence of an energetically heterogeneous surface. The occurrence of Van der Waals forces and the metal adhesion to the porosity is determinant for adsorption. The Langmuir isotherm establishes chemisorption as the basis mechanism. Chemisorption takes place in a homogeneous layer, indicating finite sites on the adsorbent's surface specific for the adsorbate [13,28]. As it was reported before, Pb^{2+} and Cd^{2+} adsorption on the biochar occurred by specific ion exchange and complexation reactions represented by Equations (13)–(15).

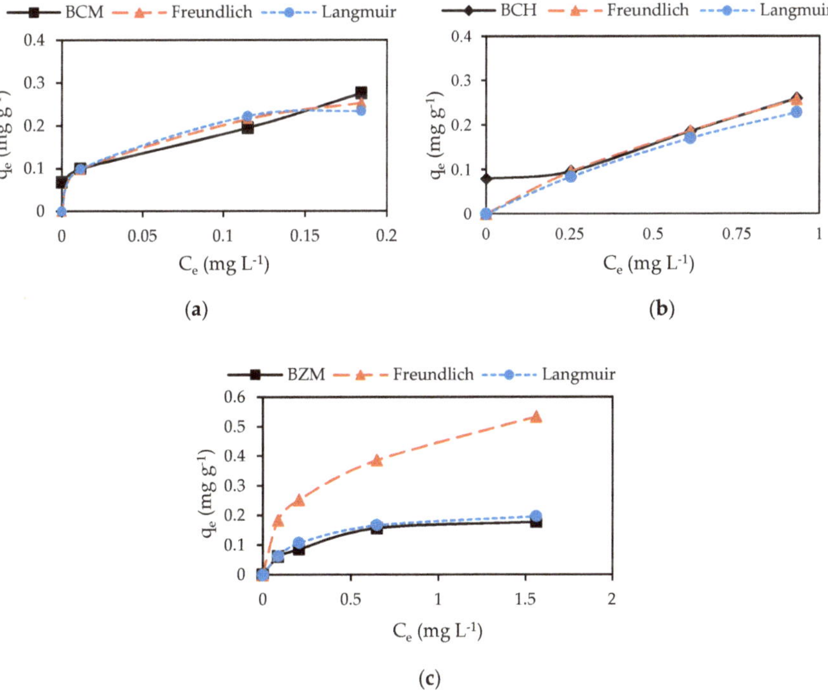

Figure 6. Adsorption Pb^{2+} isotherms: (**a**) BCM, (**b**) BCH and (**c**) BZM.

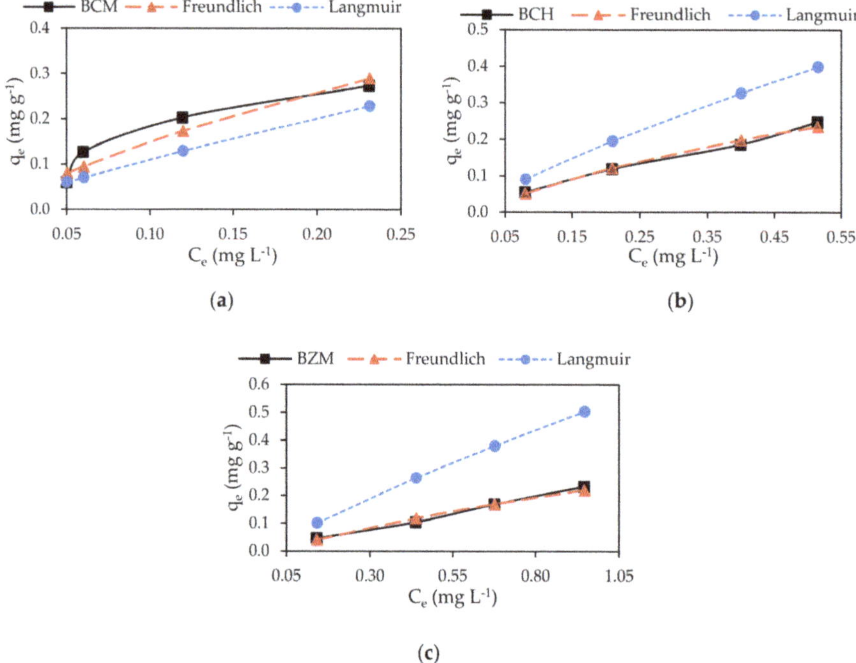

Figure 7. Adsorption Cd^{2+} isotherms: (**a**) BCM, (**b**) BCH and (**c**) BZM.

The results of this study suggest the simultaneous occurrence of physical and chemical adsorption for the Pb^{2+} and Cd^{2+} removal. Physisorption, is the mechanism that governed the Pb^{2+} and Cd^{2+} caption on biochar.

4. Conclusions

The yield of BCH $\approx$ 78.64% > CZM $\approx$ 61.23% > BCM $\approx$ 60.16%. The moisture content, the effective particle size, ash content and volatile material of biochar is in accordance with the standards for granularly activated carbon. BCM has the higher Pb^{2+} and Cd^{2+} adsorption capacity. The physicochemical characteristics of BCM are responsible for the high adsorption behavior. BCM has the highest specific surface area of biochar: BCM $\approx$ 1224 $m^2\ g^{-1}$ > BZM $\approx$ 778 $m^2\ g^{-1}$ > BCH $\approx$ 652 $m^2\ g^{-1}$. The Pb^{2+} adsorption capacities of biochar were: BCM $\approx$ 2.528 $mg\ g^{-1}$ > BCH $\approx$ 0.532 $mg\ g^{-1}$ > BZM $\approx$ 0.453 $mg\ g^{-1}$. The Cd^{2+} adsorption capacities of biochar were: BCM $\approx$ 0.314 $mg\ g^{-1}$ > BZM $\approx$ 0.155 $mg\ g^{-1}$ > BCH $\approx$ 0.049 $mg\ g^{-1}$. The experimental data were best fitted to the Freundlich isotherm model and the pseudo-second order kinetic model. The adsorption is denoted by physical mechanisms: Van der Waals forces and the biochar porosity. The adsorption on biochar is also promoted by means of chemical reactions, complexation and ion exchange.

The biochar evaluated has great potential to be used in the treatment of water, allowing Pb^{2+} and Cd^{2+} removal from drinking water. The low cost and availability of the raw material makes it an interesting proposal for water treatment. Furthermore, adsorption is a low-cost and easy-to-implement technology in treatment systems for drinking water. Although the energy requirement to obtain the biochar is important, the social benefit is also very relevant for developing countries, because with this application we are able to avoid serious diseases in children and adults, which is of evident public health interest. Furthermore, on the other hand, a sustainable use of waste is achieved, that otherwise would have to be managed, incurring significant costs.

Author Contributions: E.P.P.: Investigation, Writing—Original Draft; Writing—Review & Editing; Visualization, Project administration, Funding acquisition/D.G.: Methodology, Validation, Writing—Original Draft; Data Curation; Formal analysis; Writing—Review & Editin /C.T.: Investigation, Writing—Original Draft; Data Curation/F.O.: Conceptualization, Writing—Review & Editing, Supervision, Visualization/M.J.G.-R.: Conceptualization, Writing—Review & Editing, Supervision, Visualization. All authors have read and agreed to the published version of the manuscript.

Funding: This research was financially supported by Universidad Técnica Particular de Loja – Ecuador.

Conflicts of Interest: The authors declare no conflict of interest.

References

1. Kouassi, N.L.B.; Yao, K.M.; Trokourey, A.; Soro, M.B. Distribution, Sources, and Possible Adverse Biological Effects of Trace Metals in Surface Sediments of a Tropical Estuary. *Environ. Forens.* **2015**, *16*, 96–108. [CrossRef]
2. Chakraborty, S.; Chakraborty, P.; Nath, B.N. Lead distribution in coastal and estuarine sediments around India. *Mar. Pollut. Bull.* **2015**, *97*, 36–46. [CrossRef] [PubMed]
3. Fonseca, A.R.; Fernandes, L.F.S.; Fontaínhas-Fernandes, A.; Monteiro, S.M.; Pacheco, F.A.L. The impact of freshwater metal concentrations on the severity of histopathological changes in fish gills: A statistical perspective. *Sci. Total Environ.* **2017**, *599*, 217–226. [CrossRef] [PubMed]
4. Taiwo, A.M.; Awomeso, J.A. Assessment of trace metal concentration and health risk of artisanal gold mining activities in Ijeshaland, Osun State Nigeria—Part 1. *J. Geochem. Explor.* **2017**, *177*, 1–10. [CrossRef]
5. Li, N.; Kang, Y.; Pan, W.; Zeng, L.; Zhang, Q.; Luo, J. Concentration and transportation of heavy metals in vegetables and risk assessment of human exposure to bioaccessible heavy metals in soil near a waste-incinerator site, South China. *Sci. Total Environ.* **2015**, 144–151. [CrossRef] [PubMed]
6. Zhang, H.; Tang, Y.; Cai, D.; Liu, X.; Wang, X.; Huang, Q.; Yu, Z. Hexavalent chromium removal from aqueous solution by algal bloom residue derived activated carbon: Equilibrium and kinetic studies. *J. Hazard. Mater.* **2010**, *181*, 801–808. [CrossRef]
7. Instituto Nacional de Estadística e Informática (INEI). Estadísticas Ambientales Resultados, Info. Téc 2009. Available online: https://sinia.minam.gob.pe/documentos/informe-tecnico-estadisticas-ambientales-marzo-2009 (accessed on 2 August 2020).
8. Castillo, M.F.; Salas, E.H.; Alcantara, R.B. Estadísticas Ambientales. *Estadístic. Ambient. Info. Téc.* **2018**, *5*, 49.
9. Oviedo, A.R.; Moína-Quimí, E.; Naranjo-Morán, J.; Barcos-Arias, M. Contaminación por metales pesados en el sur del Ecuador asociada a la actividad minera. *Bionatura* **2017**, *2*, 437–441. [CrossRef]
10. Malamis, S.; Katsou, E.; Haralambous, K.J. Study of Ni(II), Cu(II), Pb(II), and Zn(II) Removal Using Sludge and Minerals Followed by MF/UF. *Water Air Soil Pollut.* **2010**, *218*, 81–92. [CrossRef]
11. Carolin, C.F.; Kumar, P.S.; Saravanan, A.; Joshiba, G.J.; Naushad, M. Efficient techniques for the removal of toxic heavy metals from aquatic environment: A review. *J. Environ. Chem. Eng.* **2017**, *5*, 2782–2799. [CrossRef]
12. Ko, D.; Lee, J.S.; Patel, H.A.; Jakobsen, M.H.; Hwang, Y.; Yavuz, C.T.; Hansen, H.C.B.; Andersen, H.R. Selective removal of heavy metal ions by disulfide linked polymer networks. *J. Hazard. Mater.* **2017**, *332*, 140–148. [CrossRef]
13. Liu, C.; Lei, X.; Wang, L.; Jia, J.; Liang, X.; Zhao, X.; Zhu, H. Investigation on the removal performances of heavy metal ions with the layer-by-layer assembled forward osmosis membranes. *Chem. Eng. J.* **2017**, *327*, 60–70. [CrossRef]
14. Visa, M. Synthesis and characterization of new zeolite materials obtained from fly ash for heavy metals removal in advanced wastewater treatment. *Powder Technol.* **2016**, *294*, 338–347. [CrossRef]
15. Bilal, M.; Rasheed, T.; Sosa-Hernández, J.E.; Raza, A.; Nabeel, F.; Iqbal, H.M. Biosorption: An Interplay between Marine Algae and Potentially Toxic Elements—A Review. *Mar. Drugs* **2018**, *16*, 65. [CrossRef] [PubMed]
16. Inyang, M.; Gao, B.; Yao, Y.; Xue, Y.; Zimmerman, A.R.; Pullammanappallil, P.; Cao, X. Removal of heavy metals from aqueous solution by biochars derived from anaerobically digested biomass. *Bioresour. Technol.* **2012**, *110*, 50–56. [CrossRef]

17. Sigdel, A.; Jung, W.; Min, B.; Lee, M.; Choi, U.; Timmes, T.; Kim, S.-J.; Kang, C.-U.; Kumar, R.; Jeon, B.-H. Concurrent removal of cadmium and benzene from aqueous solution by powdered activated carbon impregnated alginate beads. *Catena* **2017**, *148*, 101–107. [CrossRef]

18. Chen, L.; Wu, P.; Chen, M.; Lai, X.; Ahmed, Z.; Zhu, N.; Dang, Z.; Bi, Y.; Liu, T. Preparation and characterization of the eco-friendly chitosan/vermiculite biocomposite with excellent removal capacity for cadmium and lead. *Appl. Clay Sci.* **2018**, *159*, 74–82. [CrossRef]

19. Elhafez, S.A.; Hamad, H.A.; Zaatout, A.A.; Malash, G.F. Management of agricultural waste for removal of heavy metals from aqueous solution: Adsorption behaviors, adsorption mechanisms, environmental protection, and techno-economic analysis. *Environ. Sci. Pollut. Res.* **2016**, *24*, 1397–1415. [CrossRef]

20. Bouhamed, F.; Elouear, Z.; Jalel, B.; Ouddane, B. Multi-component adsorption of copper, nickel and zinc from aqueous solutions onto activated carbon prepared from date stones. *Environ. Sci. Pollut. Res.* **2015**, *23*, 15801–15806. [CrossRef]

21. Lehmann, J. A handful of carbon. *Nature* **2007**, *447*, 10–11. [CrossRef]

22. Yuan, J.-H.; Xu, R.-K.; Zhang, H. The forms of alkalis in the biochar produced from crop residues at different temperatures. *Bioresour. Technol.* **2011**, *102*, 3488–3497. [CrossRef]

23. Lafdani, E.K.; Saarela, T.; Laurén, A.; Pumpanen, J.; Palviainen, M. Purification of Forest Clear-Cut Runoff Water Using Biochar: A Meso-Scale Laboratory Column Experiment. *Water* **2020**, *12*, 478. [CrossRef]

24. Valencia, R.; Montúfar, R.; Navarrete, H.; Balslev, H. Palmas Ecuatorianas: Biología y so sostenible. In *Ecuadorian Palms: Biology and Sustainable Use*, 1st ed.; Herbario QCA de la Pontificia Universidad Católica del Ecuador: Quito, Ecuador, 2013.

25. Smith, N. Palms and People in the Amazon. In *Geobotany Studies*; Springer: Berlin, Germany, 2015.

26. Zapata, N.; Henriquez, L.; Finot, V.L. Caracterización y clasificación botánica de veintidos líneas de maní (Arachis hypogaea L.) Evaluadas en la Provincia de Nuble, Chile. *Chil. J. Agric. Anim. Sci.* **2017**, *33*, 202–212. [CrossRef]

27. Montoya, E.; Maria, F. Estudio Sobre la Comercialización de Maíz Duro en la Sierra—Centro Norte del Ecuador. Bachelor's Thesis, Universidad Politécnica Salesiana—Quito, Cuenca, Ecuador, 2010.

28. Lee, M.-E.; Park, J.H.; Chung, J.W. Comparison of the lead and copper adsorption capacities of plant source materials and their biochars. *J. Environ. Manag.* **2019**, *236*, 118–124. [CrossRef] [PubMed]

29. Vo, A.T.; Nguyen, V.P.; Ouakouak, A.; Nieva, A.; Doma, J.B.T.; Tran, H.N.; Chao, H.-P. Efficient Removal of Cr(VI) from Water by Biochar and Activated Carbon Prepared through Hydrothermal Carbonization and Pyrolysis: Adsorption-Coupled Reduction Mechanism. *Water* **2019**, *11*, 1164. [CrossRef]

30. Khiari, B.; Ghouma, I.; Ibn Ferjani, A.; Azzaz, A.A.; Jellali, S.; Limousy, L.; Jeguirim, M. Kenaf stems: Thermal characterization and conversion for biofuel and biochar production. *Fuel* **2020**, *262*, 116654. [CrossRef]

31. Lee, M.-E.; Park, J.H.; Chung, J.W. Adsorption of Pb(II) and Cu(II) by Ginkgo-Leaf-Derived Biochar Produced under Various Carbonization Temperatures and Times. *Int. J. Environ. Res. Public Health* **2017**, *14*, 1528. [CrossRef] [PubMed]

32. Liu, L.; Huang, Y.; Zhang, S.; Gong, Y.; Su, Y.; Cao, J.; Hu, H. Adsorption characteristics and mechanism of Pb(II) by agricultural waste-derived biochars produced from a pilot-scale pyrolysis system. *Waste Manag.* **2019**, *100*, 287–295. [CrossRef]

33. Colpas, F.; Taron-Dunoyer, A.A. Área superficial de carbones activados y modificados obtenidos del recurso agrícola Saccharum officinarum. *Rev. Ciencias Agric.* **2017**, *34*, 62–72. [CrossRef]

34. Park, J.; Lee, Y.; Ryu, C.; Park, Y.-K. Slow pyrolysis of rice straw: Analysis of products properties, carbon and energy yields. *Bioresour. Technol.* **2014**, *155*, 63–70. [CrossRef] [PubMed]

35. Wendlandt, W.; Dosch, E. The deltatherm V differential thermal analysis and thermogravimetry system. *Thermochim. Acta* **1987**, *117*, 45–50. [CrossRef]

36. Schlegel, M.; Ibrahim, B.; Laurel, O.; Kipping, R.; Fras, J.; Kanswohl, N.; Zosel, J. Hydrothermal Carbonization Process to Improve Transportability of Plant Biomass. *Agroproductividad* **2018**, *11*, 3–9.

37. N'Goran, K.P.D.A.; Diabaté, D.; Yao, K.M.; Kouassi, N.L.B.; Gnonsoro, U.P.; Kinimo, K.C.; Trokourey, A. Lead and cadmium removal from natural freshwater using mixed activated carbons from cashew and shea nut shells. *Arab. J. Geosci.* **2018**, *11*, 498. [CrossRef]

38. Rwiza, M.J.; Oh, S.-Y.; Kim, K.-W.; Kim, S.D. Comparative sorption isotherms and removal studies for Pb(II) by physical and thermochemical modification of low-cost agro-wastes from Tanzania. *Chemosphere* **2018**, *195*, 135–145. [CrossRef] [PubMed]

39. Colpas, F.; Tarón, A.; Fong, W. Analisis del desarrollo textural de carbones activados preparados a partir de zuro de maíz. *Temas Agrar.* **2016**, *20*, 103–112. [CrossRef]

40. Filippín, A.J.; Luna, N.S.; Pozzi, M.T.; Pérez, J.D. Obtención y Caracterización De Carbón Activado a Partir De Residuos Olivícolas Y Oleícolas Por Activacion Física Obtaining and Characterizing of Carbon Activated from Olivic and Olive-Residues By Physical Activation. *Avan. Cie. Ingeni.* **2017**, *8*, 59–71.

41. ASTM D2867-09. *Standard Test Methods for Moisture in Activated Carbon*; ASTM: West Conshohocken, PA, USA, 2014. [CrossRef]

42. ASTM D2866-11. *Standard Test Method for Total Ash Content of Activated Carbon*; ASTM: West Conshohocken, PA, USA, 2018. [CrossRef]

43. Hernández, M.; Otero-Calvis, A.; Falcón-Hernandez, J.; Yperman, Y. Características Fisicoquímicas Del Carbón Activado de Conchas de Coco Modificado Con HNO_3. *Rev. Cubana Quím.* **2017**, *29*, 26–38.

44. Ravulapalli, S.; Kunta, R. Removal of lead (II) from wastewater using active carbon of Caryota urens seeds and its embedded calcium alginate beads as adsorbents. *J. Environ. Chem. Eng.* **2018**, *6*, 4298–4309. [CrossRef]

45. Sadeek, S.A.; Negm, N.A.; Hefni, H.H.; Wahab, M.M.A. Metal adsorption by agricultural biosorbents: Adsorption isotherm, kinetic and biosorbents chemical structures. *Int. J. Biol. Macromol.* **2015**, *81*, 400–409. [CrossRef]

46. Langmuir, I. The constitution and fundamental properties of solids and liquids. Part II—Liquids. *J. Frankl. Inst.* **1917**, *184*, 721. [CrossRef]

47. Freundlich, H. Über die Adsorption in Lösungen. *Zeitschrift Physikal. Chem.* **1907**, *57*, 385–470. [CrossRef]

48. Lagergren, S. About the Theory of So-Called Adsorption of Soluble Substances. *Kungliga Svenska Vetenskapsakademiens Handlingar.* **1898**, *24*, 1–39.

49. Ho, Y.S.; McKay, G. Pseudo-Second Order Model for Sorption Processes. *Org. Process. Res. Dev.* **1998**, *21*, 866–870. [CrossRef]

50. Pinzón, M.; Vera, L. Modelamiento de la cinética de bioadsorción de Cr (III) usando cáscara de naranja kinetc modeling biosorption of Cr(III) using orange shell. *Dyna* **2009**, *76*, 95–106.

51. Castellar-Ortega, G.; Mendoza-Colina, E.D.J.; Mercado, E.R.A.; Colpas, J.E.J.; Pereira, Z.A.P.; Bravo, M.C.R. Equilibrio, cinética y termodinámica de la adsorción del colorante DB-86 sobre carbón activado de la cáscara de yuca. *Revista MVZ Córdoba* **2019**, *24*, 7231–7238. [CrossRef]

52. American Water Works Association. *AWWA B604-90 Granular Activated Carbon*; AWWA: Denver, CO, USA, 2019. [CrossRef]

53. Organización de las naciones unidas para la agricultura y la alimentación FAO. Capitulo 10. Uso Eficiente Del Carbón Vegetal. In *Métodos Simples Para Fabricar Carbón Vegetal*; FAO: Rome, Italy, 1999. Available online: http://www.fao.org/3/X5328S/X5328S05.htm (accessed on 2 May 2020).

54. Hernández, F.; Cristina, A.; Tort, S.; Recio, R. Valoración De La Calidad Del Carbón Vegetal De Las Zonas De La Efi Empresa Forestal Gran Piedra- Baconao. *Rev. Cubana Quím.* **2006**, *XVIII*, 30–38.

55. Cortés, V. Carbón. 2000. Available online: https://docplayer.es/9596492-Carbon-prof-dr-vicente-j-cortes.html (accessed on 2 May 2020).

56. Goswami, M.; Phukan, P. Enhanced adsorption of cationic dyes using sulfonic acid modified activated carbon. *J. Environ. Chem. Eng.* **2017**, *5*, 3508–3517. [CrossRef]

57. Aboua, K.N.; Yobouet, Y.A.; Yao, K.B.; Goné, D.L.; Trokourey, A. Investigation of dye adsorption onto activated carbon from the shells of Macoré fruit. *J. Environ. Manag.* **2015**, *156*, 10–14. [CrossRef]

58. Adebisi, G.A.; Chowdhury, Z.Z.; Alaba, P.A. Equilibrium, kinetic, and thermodynamic studies of lead ion and zinc ion adsorption from aqueous solution onto activated carbon prepared from palm oil mill effluent. *J. Clean. Prod.* **2017**, *148*, 958–968. [CrossRef]

59. Song, X.; Wang, L.; Zeng, Y.; Zhan, X.; Gong, J.; Li, T. Application of activated carbon modified by acetic acid in adsorption and separation of CO_2 and CH_4. Advances in energy science and environment engineering II: Proceedings of 2nd International Workshop on Advances in Energy Science and Environment Engineering. *AESEE 2018* **2018**, *1944*, 020056. [CrossRef]

60. Senthilkumar, P.; Ramalingam, S.; Sathyaselvabala, V.; Kirupha, S.D.; Sivanesan, S. Removal of copper(II) ions from aqueous solution by adsorption using cashew nut shell. *Desalination* **2011**, *266*, 63–71. [CrossRef]

61. Kumar, P.S.; Ramalingam, S.; Sathyaselvabala, V.; Kirupha, S.D.; Murugesan, A.; Sivanesan, S. Removal of cadmium(II) from aqueous solution by agricultural waste cashew nut shell. *Korean J. Chem. Eng.* **2012**, *29*, 756–768. [CrossRef]

62. Cheraghi, E.; Ameri, E.; Moheb, A. Adsorption of cadmium ions from aqueous solutions using sesame as a low-cost biosorbent: Kinetics and equilibrium studies. *Int. J. Environ. Sci. Technol.* **2015**, *12*, 2579–2592. [CrossRef]

63. Ferreira-Coelho, G.; Affonso, C.G., Jr.; Tarley, C.R.T.; Casarin, J.; Nacke, H.; Francziskowski, M.A. Removal of metal ions Cd (II), Pb (II), and Cr (III) from water by the cashew nut shell *Anacardium occidentale* L. *Ecol. Eng.* **2014**, *73*, 514–525. [CrossRef]

64. Ansari, T.M.; Hanif, M.A.; Mahmood, A.; Ijaz, U.; Khan, M.A.; Nadeem, R.; Ali, M. Immobilization of Rose Waste Biomass for Uptake of Pb(II) from Aqueous Solutions. *Biotechnol. Res. Int.* **2011**, *2011*, 685023. [CrossRef] [PubMed]

65. Ho, Y.S.; Porter, J.F.; McKay, G. Equilibrium Isotherm Studies for the Sorption of Divalent Metal Ions onto Peat: Copper, Nickel and Lead Single Component Systems. *Water Air Soil Pollut.* **2002**, *141*, 1–33. [CrossRef]

66. Albis, A.; Rangel, A.J.L.; Castilla, M.C.R. Removal of methylene blue from aqueous solutions using cassava peel (Manihot esculenta) modified with phosphoric acid//Remoción de azul de metileno de soluciones acuosas utilizando cáscara de yuca (Manihot esculenta) modificada con ácido fosfórico. *Prospectiva* **2017**, *15*, 60–73. [CrossRef]

67. Li, J.; Zheng, L.; Wang, S.-L.; Wu, Z.; Wu, W.; Niazi, N.K.; Shaheen, S.M.; Rinklebe, J.; Bolan, N.; Ok, Y.S.; et al. Sorption mechanisms of lead on silicon-rich biochar in aqueous solution: Spectroscopic investigation. *Sci. Total Environ.* **2019**, *672*, 572–582. [CrossRef] [PubMed]

68. Xu, X.; Cao, X.; Zhao, L. Comparison of rice husk- and dairy manure-derived biochars for simultaneously removing heavy metals from aqueous solutions: Role of mineral components in biochars. *Chemosphere* **2013**, *92*, 955–961. [CrossRef]

Publisher's Note: MDPI stays neutral with regard to jurisdictional claims in published maps and institutional affiliations.

© 2020 by the authors. Licensee MDPI, Basel, Switzerland. This article is an open access article distributed under the terms and conditions of the Creative Commons Attribution (CC BY) license (http://creativecommons.org/licenses/by/4.0/).

Article

Biostability of Tap Water—A Qualitative Analysis of Health Risk in the Example of Groundwater Treatment (Semi-Technical Scale)

Andżelika Domoń [1,*], Dorota Papciak [1], Barbara Tchórzewska-Cieślak [2] and Katarzyna Pietrucha-Urbanik [2]

[1] Department of Water Purification and Protection, Faculty of Civil, Environmental Engineering and Architecture, Rzeszow University of Technology, Al. Powstancow Warszawy 6, 35-959 Rzeszow, Poland; dpapciak@prz.edu.pl

[2] Department of Water Supply and Sewerage Systems, Faculty of Civil, Environmental Engineering and Architecture, Rzeszow University of Technology, Al. Powstancow Warszawy 6, 35-959 Rzeszow, Poland; cbarbara@prz.edu.pl (B.T.-C.); kpiet@prz.edu.pl (K.P.-U.)

* Correspondence: adomon@prz.edu.pl; Tel.: +48-17-865-1949

Received: 13 November 2018; Accepted: 28 November 2018; Published: 1 December 2018

Abstract: This article presents results of research which aimed to assess the impact of biofiltration processing on the biological stability of water. Effectiveness of biogenic substances removal (C, N, P) and bacteriological quality of water after the biofiltration process were discussed. The research was carried out on a semi-technical scale on natural underground water rich in organic compounds. A filter with a biologically active carbon (BAC) bed was used for the research. Despite the low water temperature of between 9–12 °C, there was a high efficiency of organic matter removal—33–70%. The number of mesophilic and psychrophilic bacteria in the water before and after the biofiltration process was comparable (0–23 CFU/mL) and met the requirements for drinking water. No *E. coli* was detected in the water samples. The biological material washed out of the filter bed did not cause deterioration of water quality which proved that the operating parameters of the biofilters were properly chosen, i.e., contact time of 30 min, filtration speed up to 3 m/h. Reduction of the content of nutrients in the treated water limits the risk of microbial growth and thus the emergence of biological growth in the distribution system.

Keywords: bacteriological contamination of water; biofiltration; biological stability; water treatment

1. Introduction

The combination of unit processes used in water treatment technological systems should ensure not only removal of pollutants present in normative values but also guarantee water quality that reduces the risk of secondary contamination during transport to the recipient [1].

It is believed that the main cause of changes in the quality composition of water during its transport is, apart from the technical condition of the network, water instability. The result of instability is the intensification of corrosive processes through the precipitation and dissolution of accumulated deposits and the formation of biofilms [2,3]. All the factors enabling the creation and development of biofilms in water distribution systems are not presently sufficiently known [4]. A water that does not contain microorganisms and does not sustain their development in the water supply network is considered to be biologically stable. The striving to achieve the biological stability of water is associated with the need to ensure an extremely low content of food substrates, especially biodegradable organic carbon (BDOC), easily digestible nitrogen, and phosphorus [5–9].

The unsatisfactory efficiency of biogens removal has led to the need to expand conventional technological systems with a biofiltration process in which contaminants are removed by sorption, assimilation, and biodegradation [10–12].

Biofiltration makes it possible to eliminate the basic nutrients that ensure the growth of microorganisms and precursors of disinfection by-products (this is demonstrated by the decrease in UV absorbance) [13–15].

However, it remains doubtful how the efficiency of the biofiltration process will change after exhausting the adsorptive capacity of activated carbon. Will the water after the biofiltration process still be safe in sanitary terms and how will it affect the stability of tap water?

The greatest concerns relate to the possibility of microorganisms or fragments of biofilms, formed on activated carbon seeds which fill the biofilter, getting into the treated water. The fact that the final disinfection process does not completely eliminate bacteria, viruses, and fungi present in the treated water further intensifies these fears [16,17].

The aim of the research was to assess the impact of the biofiltration process on the microbiological quality of water and the effectiveness of biogenic substances removal which determine the biological stability of tap water.

2. Materials and Methods

2.1. Characteristics of Drawn Water (Raw)

The Water Treatment Plant (WTP) is supplied from an unconfined quaternary aquifer with a depth of about 15 m bgl using 27 wells. The physicochemical quality of the water being captured is presented in Table 1.

Table 1. Physicochemical quality of underground water (data from WTP).

Indicator	Unit	Range of Changes
Total Organic carbon (TOC)	$g\,O_2/m^3$	11.0–14.5
Permanganate value	$g\,O_2/m^3$	11.0–18.1
Turbidity	NTU	8.0–14.0
Colour	$g\,Pt/m^3$	40–100
Ammonia nitrogen	$g\,NH_4^+/m^3$	1.20–1.98
pH	-	6.4–7.0
Temperature	°C	10.8–12.1
Hardness	$g\,CaCO_3/m^3$	200–470
Sulfate	$g\,SO_4^{2-}/m^3$	60–240
Conductivity	mS/cm	430–1016
Alkalinity	val/m^3	2.5–4.5

2.2. Technology System for Water Treatment

The drawn water is directed to a collective well (stoppage time of 2–4 h, depending on the current water production). The water is then directed to a cascade of oxygenation. A dispensing potassium permanganate chemical-oxidant with a dose of 2.1–2.4 mg/L and a PAX-18 coagulant in the amount of 120–140 mg/L is situated just below the cascade (11.0–12.6 mg Al/L). Vertical coagulation–sedimentation chambers with a contact time of 6–8 h are the next stage of the water treatment. Lime milk is dosed (about 10.0 mg/L) to the water in the chambers, which is directed to horizontal settler raising the pH. The retention time in the settlers is several hours. The next step involves filtration of water at a speed of 1.5–3 m/h and the final disinfection with sodium hypochlorite (Figure 1).

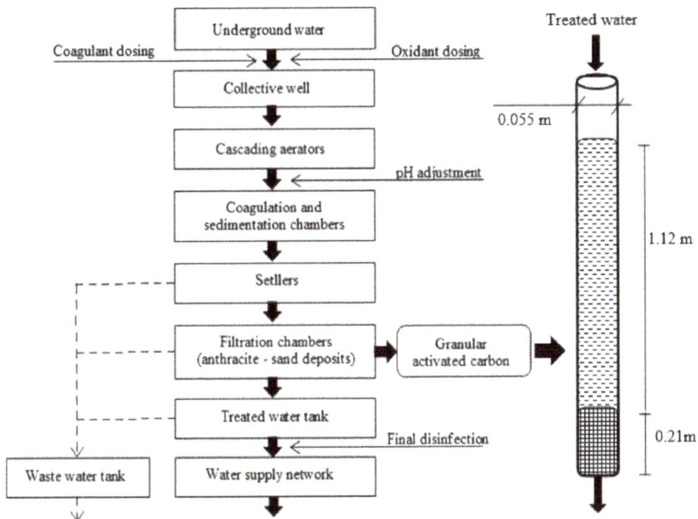

Figure 1. Technological flowchart for the Water Treatment Plant.

The filter used was a biologically active deposit of activated carbon with the parameters contained in Table 2.

Table 2. Parameters of the biofilter model [18].

Operating Parameters	Value	Properties of the Filter Material	Value
Height of the carbon bed, m	1.12	Granulation, mm	1–4
Diameter, m	0.055	Specific surface, m^2/g	950–1050
Filtration velocity, m/h	1.5–2.0	The iodine value, mg/g	998
Contact time, h	0.5	pH aqueous extract	11

2.3. Research Procedure

The research was carried out over a period of one calendar year. The list of controlled water parameters and analysis procedures is presented in Table 3. The obtained results were subjected to statistical analysis with the use of STATISTICA12 software.

2.4. Evaluation of Work of Model Filter-Test EMS

The development of activity within biosorption beds, and the observations regarding the relationships between sorption and biodecomposition processes, were recorded based on the Eberhardt, Madsen, Sontheimer (EMS) test [19–21]. The EMS test is based on the value of the indicator, described by the relation between the change in COD or permanganate index and the loss of dissolved oxygen, taking place during filtration.

$$S = \Delta COD / \Delta O_2 \tag{1}$$

where:

ΔCOD—loss of COD (with $K_2Cr_2O_7$ or with $KMnO_4$)
ΔO_2—loss of dissolved oxygen

When:

- $S = 1$ adsorption and biodecomposition happen with identical intensiveness,

- S > 1 adsorption dominates,
- S < 1 biodecomposition dominates,
- S = 0, ΔCOD = 0, ΔO_2 > 0 sorption and biodecomposition processes stopped
- S undetermined, ΔCOD > 0, ΔO_2 = 0 sorption present, biodecomposition absent ΔCOD = 0, ΔO_2 = 0 sorption and biodecomposition absent

The presence of the biofilm was confirmed by observation, using a light microscope and an electron scanning microscope JOEL SEM 5500 LV (JEOL Ltd., Tokyo, Japan).

Table 3. Summary of analytical methods for laboratory experiments [22].

Parameters	Analytical Method/Standard
Total organic carbon (TOC)	TOC analyzer Sievers 5310 C (SUEZ, Boulder, CO, USA)
Permanganate value	The permanganate method
UV absorbance	Spectrophotometric method
Inorganic nitrogen content ($N\text{-}NH_4^+ + N\text{-}NO_2^- + N\text{-}NO_3^-$)	$N\text{-}NH_4^+$: direct nessleryization method using Merck spectrophotometer $N\text{-}NO_2^-$: colorimetric method by Nitrite Test Merck 1.14408 $N\text{-}NO_3^-$: spectrophotometric method with sodium salicylate
Inorganic phosphorus content	Spectrophotometric method with ammonium molybdate using Merck spectrophotometer
Dissolved oxygen (DO)	Electrochemical method using a Hach Lange oxygen probe
The total number of bacteria at 37 °C after 24 h (mesophilic bacteria)	Traditional culture method using A Agar from BTL Ltd.
The total number of bacteria at 22 °C after 72 h (psychrophilic bacteria)	
Escherichia coli bacteria	Membrane filtration procedure using Endo agar

3. Results

Expanding a conventional treatment system with a biofiltration process on granulated active carbon resulted in an improvement of physicochemical parameters (Figure 2, Figure 3 and Figure 9) and did not cause the deterioration of the treated water's microbiological parameters (Figures 7 and 8).

The average monthly values of total organic carbon in the water feeding the carbon filter ranged from 8.13 to 11.61 mg C/dm^3, whereas, after the biofiltration process, the values ranged from 3.47 to 6.31 mg C/dm^3 (Figure 2). The highest efficiency of removing organic compounds, by 70%, was observed in February. From April to November, the effectiveness of total organic carbon (TOC) removal was relatively stable at 33%.

The average efficiency of absorbance reduction after the biofiltration process ranged from 4% to 72%, and the lowest obtained value was 3.40 m^{-1} (Figure 2). The analysis of obtained test results showed a strong relationship between the concentration of TOC and UV absorbance, for which the Pearson correlation coefficient reached a value of R = 0.77. The measurement of the total organic carbon content at many water treatment plants is often replaced by the ultraviolet absorbance test.

The concentration of organic matter expressed by oxidation fluctuated around the recommended value (5 mg O_2/dm^3), while the introduction of a biofiltration process allowed lowering of the average monthly values of this parameter by a factor of two (Figure 3).

The increase in the efficiency of organic matter removal in the filtration process through a bed of activated carbon was mainly caused by the process of biodegradation occurring with the participation of microorganisms developing on the surface of grains [19]. The dominant biosorption process was demonstrated by the reduction of oxidation and oxygen depletion (Figures 3 and 4).

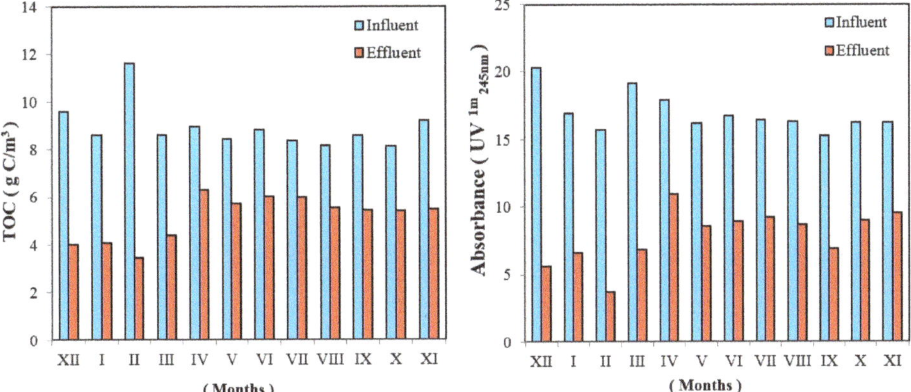

Figure 2. Average monthly values of total organic carbon and UV absorbance in water before and after the biofiltration process.

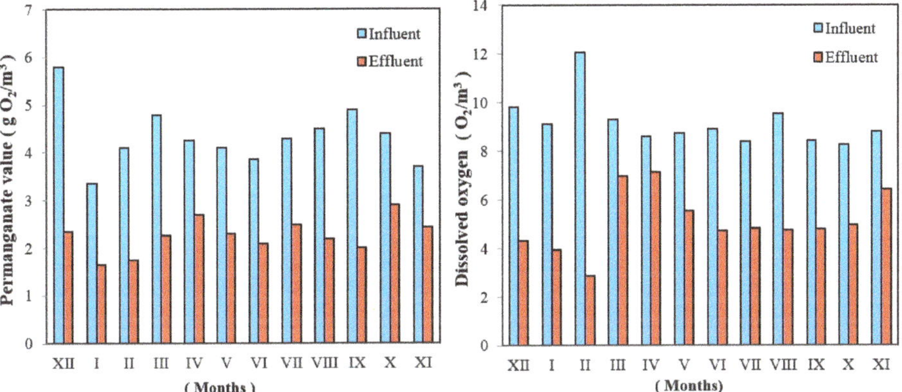

Figure 3. Average monthly values of permanganate value and dissolved oxygen in water before and after the biofiltration process.

The nature of the carbon bed's work and its biological activity was determined using the Eberhardta, Madsena, and Sointheimera (EMS test). The average values of the S index obtained for the one-year study are shown in Figure 4.

The EMS test confirmed that the biodegradation process dominates on the surface of the activated carbon. S index values were in the range from 0.26 to 0.8. The exception was in the months of March and April, during which the increase of S index was above 1, indicating the dominance of the sorption process over biodegradation.

The presence of biofilm on the surface of activated carbon is confirmed by images from light (Figure 5) and scanning microscopes (Figure 6).

The average number of mesophilic and psychrophilic bacteria in the water before and after the process was at a comparable level, i.e., 0–18 CFU/mL in the case of mesophilic and 0–23 CFU/mL—psychrophilic (Figure 7). The presence of *E. coli* was not detected in the water samples after the biofiltration process during the entire test period. The presence of *E. coli* in an amount of 1–3 CFU/mL constitutes low, 4–6 CFU/mL medium, and 7–9 CFU/mL a high health risk [23]

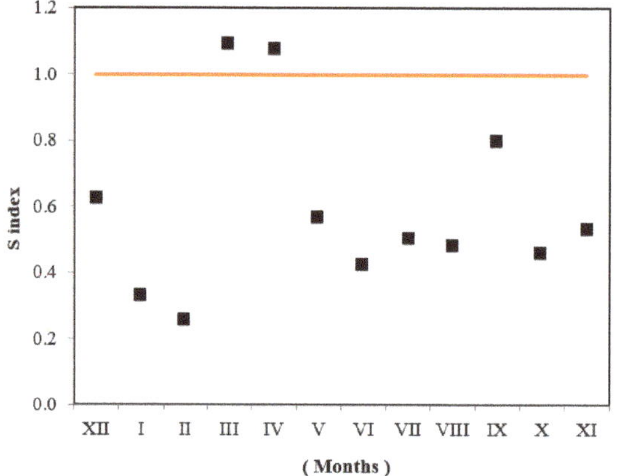

$$S = \frac{\Delta\, COD}{\Delta\, DO}$$

S > 1 adsorption dominates
S < 1 biodegradation dominates,
S = 1 adsorption and biodegradation
occurs with the same intensity

Figure 4. Average monthly changes in S indicator after biofiltration process.

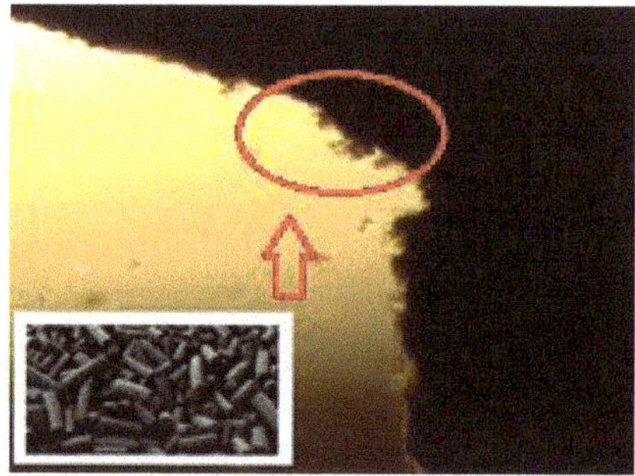

Figure 5. Edge of granular activated carbon grain covered with biological membrane (photos from a light microscope).

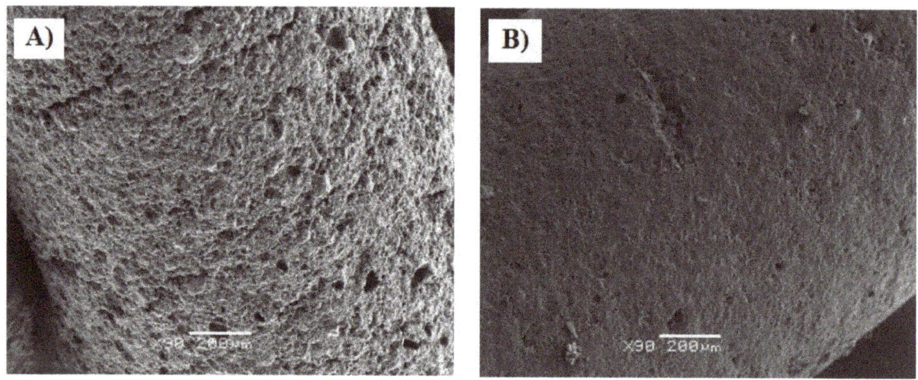

Figure 6. Biological membrane formed on granular activated carbon ((**A**): before biosorption, (**B**): after biosorption))—photos from a scanning microscope.

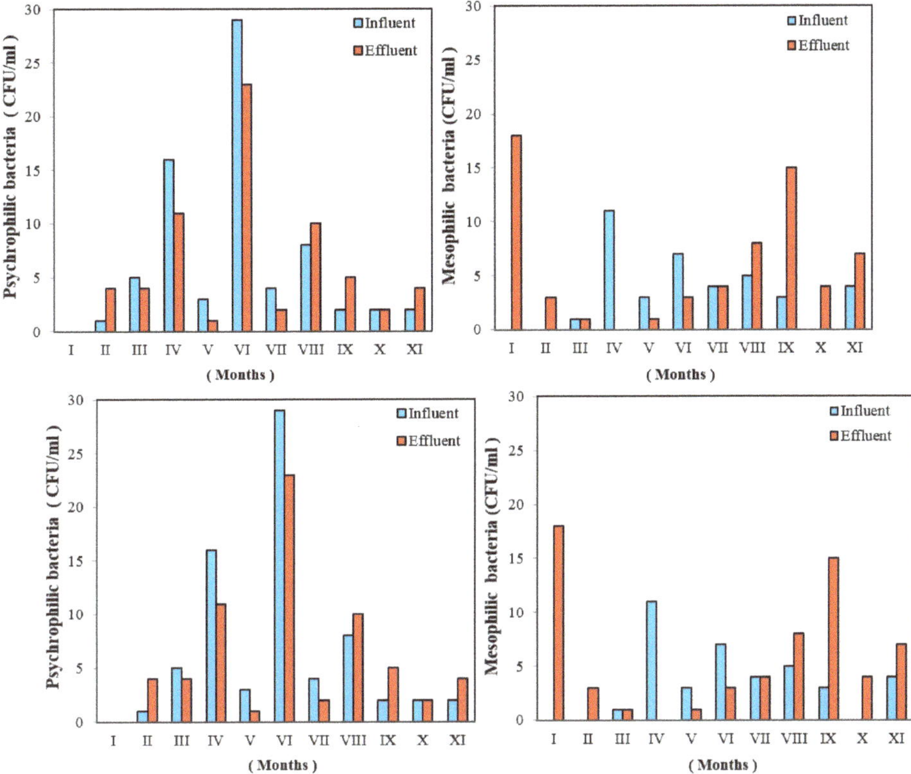

Figure 7. Total number of mesophilic and psychrophilic bacteria in water before and after the biofiltration process.

Introduction of the biofiltration process allowed obtaining of water that meets the microbiological requirements as defined by Council Directive 98/83/EC of 3 November 1998 on the quality of water intended for human consumption. In addition, the quality of treated water corresponded to the requirements for bottled water (bottle sales), as the number of mesophilic and psychrophilic bacteria did not exceed the normative values of 20 and 100 CFU/mL, respectively (Figure 7) [24].

The statistical analysis of the microbiological research results (Figure 8) fully confirms that a properly conducted biofiltration process is safe. The biological material washed out of the filter bed will not affect the sanitary condition of water if the biofilter parameters are optimal for the biodegradation process, i.e., contact time (30 min), filtration speed (1–3 m/h).

Figure 9 compares the physicochemical quality of water before and after the biofiltration process on granulated activated carbon. Reducing the content of nutrients in water limits the risk of loss of biological stability of water entering the distribution system. The threshold values of parameters limiting secondary development of microorganisms in distribution systems should be lower than 0.25 g C/m^3 BDOC, 0.2 g N$_{norg}$/m^3 and 0.03 g PO$_4^{3-}$/m^3 [25].

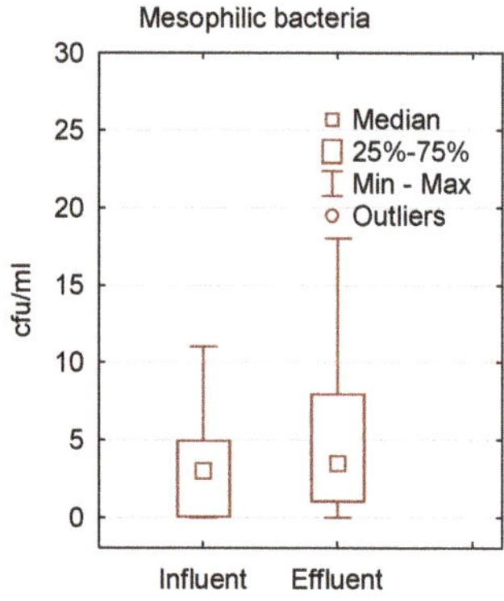
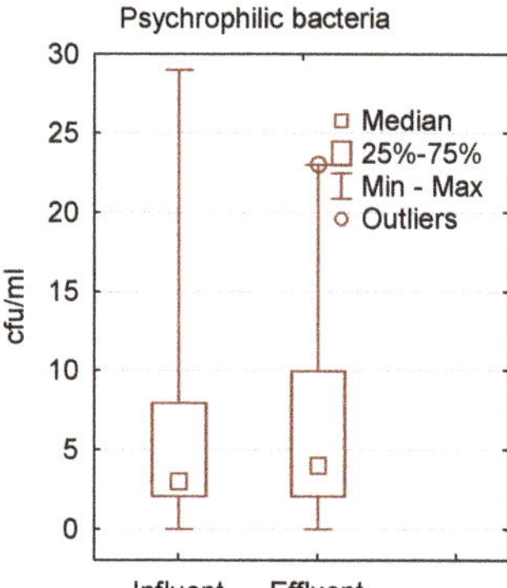

Figure 8. Statistical analysis of microbiological results obtained for water before and after biofiltration process on granular activated carbons.

For the whole of the research period, the average content of inorganic nitrogen in water reaching the biofilter was 0.76 g N_{norg}/m^3, whereas after filtration through the active carbon bed it was 0.62 g N_{norg}/m^3. The value recommended for maintaining biostability—0.2 g N_{norg}/m^3—was not obtained (Figure 10). The content of phosphate ions varied in the range of 0–0.0035 g PO_4^{3-}/m^3 and met the requirements necessary to maintain water biostability (0.03 gPO_4^{3-}/m^3) (Figure 10).

Analyzing the three-level scale of risk, i.e., tolerated, controlled, and non-acceptable risk, it was found that in a conventional treatment system, only 4.7% of the samples tested are in the tolerable risk zone, while the biofiltration process increases this value to 28.6%. The remaining samples correspond to the level of safety requiring control and reduction, in which there are reasons to maintain water biostability [14].

The comparison of median of the treated water's physicochemical parameters in a conventional system and in the system extended by the biofiltration process indicates the justifiability of introducing biofiltration using granulated activated carbon. Reducing the content of nutrients in treated water reduces the risk of microbial growth and thus the emergence of biological growth in the distribution system.

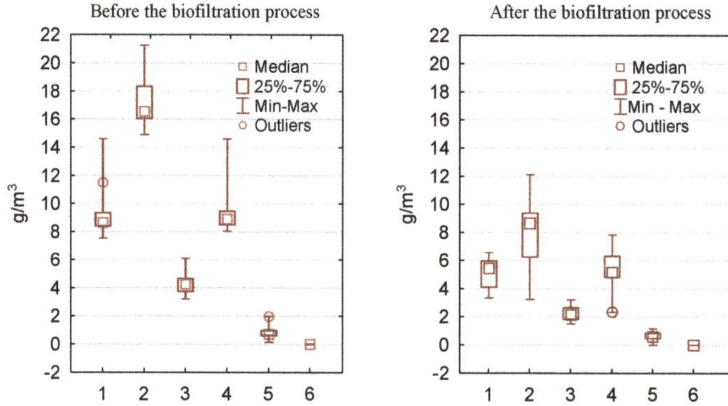

Figure 9. Statistical analysis of selected physicochemical results obtained for water before and after the biofiltration process on granular activated carbon deposits. Designations: (1) TOC (g C/m^3); (2) UV absorbance (UV^{1m}_{254}); (3) Permanganate value (g O_2/m^3); (4) Dissolved oxygen (g O_2/m^3); (5) Total inorganic nitrogen (g N_{norg}/m^3); (6) Total phosphorus (g PO_4^{3-}/m^3); (N = 44).

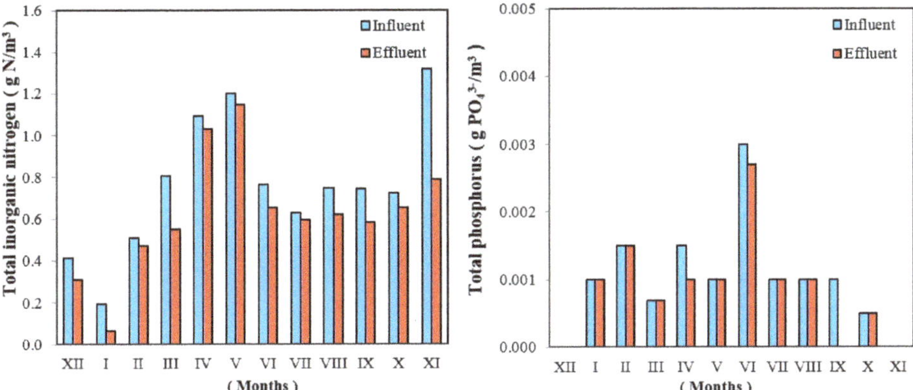

Figure 10. Average monthly values of total inorganic nitrogen and total phosphorus in water before and after the biofiltration process.

4. Discussion

The effectiveness of removing organic pollutants on active carbon beds has been confirmed by numerous studies [15,19,26–29]. Conventional water treatment systems eliminate natural organic substances up to 40% [11,21,24], while systems extended by biofiltration, depending on the time of biofilter exploitation, achieve from 24–100% [15,19,30–32]. The effectiveness of organic matter removal in metabolic processes is lower (24–42%) than in the adsorption process (70–100%), but the effectiveness of biofilter operation may exceed 10 years [10,33].

The research conducted for the first operational period of the biofilter [19] showed that the combination of sorption processes with biodegradation until exhaustion of the capacity of the carbon bed allowed for 100% removal of organic matter. Creation of biofilm before exhaustion of the sorption capacity of carbon allowed extending of the filtration cycle but the effectiveness of removing TOC decreased to 70% [19].

The results of research presented in this paper indicate that after a year of using a biofilter the efficiency of organic matter removal decreased by 50%. The observed decrease in the effectiveness of the biofilter could be caused by the depletion of the sorption capacity of carbon and the decrease in the activity of microorganisms inhabiting the carbon bed.

The water temperature is an important parameter determining the activity of microorganisms. Moll et al. found that the removal efficiency of dissolved organic carbon at 20 °C was 24%, and at 5 °C the efficiency was reduced to 15% [34]. Emelko et al. observed similar differences in the efficiency of the biodegradation process—92% for 21–25 °C and 58% for 1–3 °C [35]. Selbes et al. found seasonal variability of biofiltration efficiency. The effectiveness of removing dissolved organic carbon was lower in cooler months (~5%) and higher in warmer months (~24%) [36]. Halle et al. also confirm that the supply water temperature affects the efficiency of biofiltration [37]. Low temperature not only reduces the rate of metabolization of the substrate but also affects the structure and diversity of the biological membrane [34,38]. Hallam determined that the microbial activity at 7 °C is 50% lower than at 17 °C [3].

In the analyzed case, the problem concerned removal of organic matter from groundwater, which, unlike surface water, is characterized by low temperature variability throughout the year. The temperature could not constitute the main reason for reducing the activity of microorganisms because the highest efficiency of biofilters was observed in the winter months, i.e., December–March.

Literature indicates that the microbial activity correlates with the availability of nutrients, i.e., the content of carbon, nitrogen, and phosphorus in the water flowing into the biosorption field. The C:N:P ratio should be 100:10:1 and deficiency of any of the constituents limits the growth and development of microorganisms, thus interfering with the course of the biodegradation process [39,40]. Dhawan et al. controlled the effectiveness of three biofilters in which the treated water differed in terms of nutrient content as follows: 546:24:1, 100:10:1, and 25:10:1. The Dissolved Organic Carbon (DOC) removal efficiency was 23.5%, 28.6%, and 33.5% respectively [40]. In the case of waters containing natural organic matter and inorganic nitrogen, phosphate ions are essential [41,42]. Their insufficient content inhibits the growth of microorganisms to a much greater extent than is the case for other biogens [6,43]. Phosphorus concentrations below 0.005 mg/dm^3 inhibit the transport of phosphorus to bacterial cells [44].

The water flowing into the biofilter was characterized by a low phosphorus content (0–0.0035 g PO$_4$$^{3-}$/m^3) which could have influenced the activity of microorganisms and the efficiency of the biofiltration process. The observed lack of changes in the content of phosphorus in water after biofiltration with the simultaneous removal of nitrogen (18%) and TOC (33%) could suggest the existence of another source of phosphorus. It is likely that the carbon that fills the biofilter has been chemically activated with phosphorus-containing compounds.

It should be noted that despite the unfavorable relationship between the content of biogenic substances (C:N:P, 28500:217:1), there has been a smooth transition of the sorption process into biodegradation. The only explanation for the effective operation of the biofilter in the initial period of exploitation could be a method of coal grains' chemical activation. Perhaps supplementing the phosphorus content in an amount that would allow the optimal biogen ratio would help to increase the efficiency of natural organic matter removal. The decisive role of phosphorus should be kept in mind when water directed to a biofilter is subjected to coagulation, effectively eliminating phosphorus from the treated water.

The effectiveness of removing impurities in the BAC process also depended on the activity and type of microorganisms forming the biological membrane [33,45]. Studies on the biological activity of carbon bed samples were presented, among others, in the thesis [28,46–48]. Literature data show that the biomass concentration demonstrates vertical stratification and decreases with the depth of the bed, however [45,47,49–52], this is not the rule [53]. Differences in the composition and concentration of biomass depend on the availability of nutrients and the concentration of dissolved oxygen [54]. Less attention has been paid to the microbiological quality of the filtrate. In the works [55,56], small changes were shown between the quality of water directed to biofilters and flowing from the biofilter (the study was conducted for surface water treated with ozonation before biofiltration).

When analyzing the microbiological quality, it was found that the number of mesophiles and psychrophiles in water before and after biofiltration was at a comparable level from 0 to 23 CFU/ mL. The biological material washed out of the filter bed did not cause deterioration of water quality which proved that the operating parameters of the biofilters were properly chosen, i.e., contact time (30 min), filtration speed (up to 3 m/h).

5. Conclusions

The biofiltration process improves the quality and biological stability of treated water, reducing the health risk and risks associated with threats to the technical infrastructure.

The number of mesophilic and psychrophilic bacteria in the water before and after the biofiltration process was comparable (0–23 CFU/mL) and met the requirements for drinking water. No *E. coli* was detected in the water samples. The biological material washed out of the filter bed did not cause deterioration of the water quality which proved that the operating parameters of the biofilters were properly chosen.

The efficiency of TOC removal in the biofiltration process ranged from 70% to 33%. The reason for the decrease in the effectiveness of the process of organic matter removal in the second year of biofilter exploitation could be depletion of the adsorption capacity of the carbon bed and reduction of the activity of microorganisms colonizing the bed. Too low content of phosphorus in the water entering the biofilter could affect the efficiency of the biodegradation and assimilation process.

The values of the median of treated water's physicochemical and bacteriological parameters in the system extended by the biofiltration process prove the justifiability of introducing this process into the technological system of water treatment.

Author Contributions: All authors equal contributed to the manuscript.

Funding: This research was funded by Faculty of Civil and Environmental Engineering and Architecture, Rzeszow University of Technology, 35-959 Rzeszow, Poland.

Conflicts of Interest: The authors declare no conflict of interest.

References

1. Vreeburg, J.H.G.; Boxall, J.B. Discolouration in potable water distribution systems: A review. *Water Res.* **2007**, *41*, 519–529. [CrossRef] [PubMed]
2. Srinivasan, S.; Harrington, G.W. Biostability analysis for drinking water distribution systems. *Water Res.* **2007**, *41*, 2127–2138. [CrossRef] [PubMed]
3. Hallam, N.; West, J.; Forster, C.; Simms, J. The potential for biofilm growth in water distribution systems. *Water Res.* **2001**, *35*, 4063–4071. [CrossRef]
4. Manuel, C.M.; Nunes, O.C.; Melo, L.F. Dynamics of drinking water biofilm in flow/non-flow conditions. *Water Res.* **2007**, *41*, 551–562. [CrossRef] [PubMed]
5. Volk, C.J.; LeChevallier, M.W. Assessing biodegradable organic matter. *Am. Water Works Assoc.* **2000**, *92*, 64–76. [CrossRef]
6. Chu, C.; Lu, C.; Lee, C. Effects of inorganic nutrients on the regrowth of heterotrophic bacteria in drinking water distribution systems. *J. Environ. Manag.* **2005**, *74*, 255–263. [CrossRef] [PubMed]
7. Liu, X.; Wang, J.; Liu, T.; Kong, W.; He, X.; Jin, Y.; Zhang, B. Effects of assimilable organic carbon and free chlorine on bacterial growth in drinking water. *PLoS ONE* **2015**, *10*. [CrossRef] [PubMed]
8. Escobar, I.C.; Randall, A.A. Assimilable organic carbon (AOC) and biodegradable dissolved organic carbon (BDOC). *Water Res.* **2001**, *35*, 4444–4454. [CrossRef]
9. Van Der Kooij, D. Biological stability: A multidimensional quality aspect of treated water. *Water Air. Soil Pollut.* **2000**, *123*, 25–34. [CrossRef]
10. Simpson, D.R. Biofilm processes in biologically active carbon water purification. *Water Res.* **2008**, *42*, 2839–2848. [CrossRef] [PubMed]
11. Kaleta, J.; Kida, M.; Koszelnik, P.; Papciak, D.; Puszkarewicz, A.; Tchórzewska-Cieślak, B. The use of activated carbons for removing organic matter from groundwater. *Arch. Environ. Prot.* **2017**, *43*, 32–41. [CrossRef]
12. Rattier, M.; Reungoat, J.; Gernjak, W.; Keller, J. *Organic Micropollutant Removal by Biological Activated Carbon Filtration: A Review*; Urban Water Security Research Alliance-University of Queensland: Brisbane, QLD, Australia, 2012.
13. Korshin, G.V.; Wu, W.W.; Benjamin, M.M.; Hemingway, O. Correlations between differential absorbance and the formation of individual DBPs. *Water Res.* **2002**, *36*, 3273–3282. [CrossRef]
14. Tchórzewska-Cieślak, B.; Papciak, D.; Pietrucha-Urbanik, K.; Pietrzyk, A. Safety analysis of tap water biostability. *Archit. Civ. Eng. Environ.* **2018**, *11*, 149–154. [CrossRef]
15. Liu, W.; Wu, H.; Wang, Z.; Ong, S.; Hu, J.; Ng, W. Investigation of assimilable organic carbon (AOC) and bacterial regrowth in drinking water distribution system. *Water Res.* **2002**, *36*, 891–898. [CrossRef]
16. Camper, A.K.; LeChevallier, M.W.; Broadaway, S.C.; McFeters, G.A. Bacteria associated with granular activated carbon particles in drinking water. *Appl Environ. Microbiol.* **1986**, *52*, 434–438.
17. Wang, Q.; You, W.; Li, X.; Yang, Y.; Liu, L. Seasonal changes in the invertebrate community of granular activated carbon filters and control technologies. *Water Res.* **2014**, *51*, 216–227. [CrossRef] [PubMed]
18. Pietrzyk, A.; Papciak, D. Removal of organic compounds from natural underground water in sorption and sono-sorption processes on selected activated carbons. *E3S Web Conf.* **2017**, *17*. [CrossRef]
19. Papciak, D.; Kaleta, J.; Puszkarewicz, A.; Tchórzewska-Cieślak, B. The use of biofiltration process to remove organic matter from grounwater. *J. Ecol. Eng.* **2016**, *17*, 119–124. [CrossRef]
20. Wolborska, A.; Zarzycki, R.; Cyran, J.; Grabowska, H.; Wybór, M. Evaluation of the biological activity of carbon filters in surface water treatment on the example of a water supply "Sulejów-Łódź". *Ochr. Śr.* **2003**, *25*, 27–32. (In Polish)

21. Perchuć, M.; Grabińska-Łoniewska, A. *Technology Research the Effect of the Type of Coal Used in the BAF to Remove Humic Acids from Drinking Water*; Czestochowa University of Technology: Częstochowa, Poland, 1998; pp. 144–153. (In Polish)

22. Rice, E.W.; Baird, R.B.; Eaton, A.D. *Standard Methods for the Examination of Water and Wastewater*, 23rd ed.; American Public Health Association, American Water Works Association, Water Environment Federation: Washington, DC, USA, 2017.

23. World Health Organization. *Guidelines for Drinking-Water Quality*, 4th ed.; World Health Organization: Geneva, Switzerland, 2011.

24. Council Directive 98/83/EC of 3 November 1998 on the Quality of Water Intended for Human Consumption. 1998. EUR-Lex Web site. Available online: https://eur-lex.europa.eu/legal-content/EN/TXT/?uri=CELEX%3A31998L0083 (accessed on 29 November 2018).

25. Wolska, M. *Removal of Biogenic Substances in the Technology of Water Purification Intended for Human Consumption*; Oficyna Wydawnicza Politechniki Wrocławskiej: Wrocław, Poland, 2015. (In Polish)

26. Hijnen, W.; Schurer, R.; Martijn, B.; Bahlman, J.A.; Hoogenboezem, W.; van der Wielen, P. Removal of Easily and more Complex Biodegradable NOM by Full-Scale BAC Filters to Produce Biological Stable Drinking Water. In *Progressing in Slow Sand and Alternative Biofiltration Processes—Further Developments and Applications*; Nakamoto, N., Graham, N., Collins, M.R., Eds.; IWA Publishing: London, UK, 2014; pp. 1–8.

27. Urbanowska, A.; Kabsch-Korbutowicz, M. Characteristics of natural organic matter removed from water along with its treatment. *Environ. Prot. Eng.* **2016**, *42*, 183–195. [CrossRef]

28. Seredyńska-Sobecka, B.; Tomaszewska, M.; Janus, M.; Morawski, A.W. Biological activation of carbon filters. *Water Res.* **2006**, *40*, 355–363. [CrossRef] [PubMed]

29. Korotta-Gamage, S.M.; Sathasivan, A. A review: Potential and challenges of biologically activated carbon to remove natural organic matter in drinking water purification process. *Chemosphere* **2017**, *167*, 120–138. [CrossRef] [PubMed]

30. Pietrzyk, A.; Papciak, D. The effectiveness of organic matter removal in unit processes of the technological groundwater treatment system. *E3S Web Conf.* **2018**, *44*. [CrossRef]

31. Lohwacharin, J.; Yang, Y.; Watanabe, N.; Phetrak, A.; Sakai, H.; Murakami, M.; Oguma, K.; Takizawa, S. Characterization of DOM Removal by Full-Scale Biological Activated Carbon (BAC) Filters Having Different Ages. In Proceedings of the IWA Specialty Conference on Natural Organic Matter, Costa Mesa, CA, USA, 27–29 July 2011; pp. 1–15.

32. Rigobello, E.S.; Dantas, A.D.B.; Di Bernardo, L.; Vieira, E.M. Removal of diclofenac by conventional drinking water treatment processes and granular activated carbon filtration. *Chemosphere* **2013**, *92*, 184–191. [CrossRef] [PubMed]

33. Liao, X.; Chen, C.; Chang, C.-H.; Wang, Z.; Zhang, X.; Xie, S. Heterogeneity of microbial community structures inside the up-flow biological activated carbon (BAC) filters for the treatment of drinking water. *Biotechnol. Bioprocess Eng.* **2012**, *17*, 881–886. [CrossRef]

34. Moll, D.M.; Summers, R.S.; Fonseca, A.C.; Matheis, W. Impact of temperature on drinking water biofilter performance and microbial community structure. *Environ. Sci. Technol.* **1999**, *33*, 2377–2382. [CrossRef]

35. Emelko, M.B.; Huck, P.M.; Coffey, B.M.; Smith, E.F. Effects of media, backwash, and temperature on full-scale biological filtration. *J. Am. Water Works Assoc.* **2006**, *98*, 61–73. [CrossRef]

36. Selbes, M.; Amburgey, J.; Peeler, C.; Alansari, A.; Karanfil, T. Evaluation of seasonal performance of conventional and phosphate-amended biofilters. *J. Am. Water Works Assoc.* **2016**, *108*, 523–532. [CrossRef]

37. Hallé, C.; Huck, P.M.; Peldszus, S. Emerging contaminant removal by biofiltration: Temperature, concentration, and EBCT impacts. *J. Am. Water Works Assoc.* **2015**, *107*, 364–379. [CrossRef]

38. Corre, C.; Couriol, C.; Amrane, A.; Dumont, E.; Andrès, Y.; Le Cloirec, P. Efficiency of biological activator formulated material (BAFM) for volatile organic compounds removal—preliminary batch culture tests with activated sludge. *Environ. Technol.* **2012**, *33*, 1671–1676. [CrossRef] [PubMed]

39. Lauderdale, C.; Chadik, P.; Kirisits, M.J.; Brown, J. Engineered biofiltration: Enhanced biofilter performance through nutrient and peroxide addition. *J. Am. Water Works Assoc.* **2012**, *104*, 298–309. [CrossRef]

40. Dhawan, S.; Basu, O.D.; Banihashemi, B. Influence of nutrient supplementation on DOC removal in drinking water biofilters. *Water Sci. Technol. Water Supply* **2017**, *17*, 422–432. [CrossRef]

41. Lehtola, M.J.; Miettinen, I.T.; Lampola, T.; Hirvonen, A.; Vartiainen, T.; Martikainen, P.J. Pipeline materials modify the effectiveness of disinfectants in drinking water distribution systems. *Water Res.* **2005**, *39*, 1962–1971. [CrossRef] [PubMed]
42. Lehtola, M.J.; Miettinen, I.T.; Keinänen, M.M.; Kekki, T.K.; Laine, O.; Hirvonen, A.; Vartiainen, T.; Martikainen, P.J. Microbiology, chemistry and biofilm development in a pilot drinking water distribution system with copper and plastic pipes. *Water Res.* **2004**, *38*, 3769–3779. [CrossRef] [PubMed]
43. Lehtola, M. Microbially available organic carbon, phosphorus, and microbial growth in ozonated drinking water. *Water Res.* **2001**, *35*, 1635–1640. [CrossRef]
44. Rosenberg, H. Phosphate Transport in Prokaryotes. In *Ion Transport in Prokaryotes*; Rosen, B.P., Ed.; Academic Press: Cambridge, MA, USA, 1987; pp. 205–248.
45. Ko, Y.-S.; Lee, Y.-J.; Nam, S. Evaluation of a pilot scale dual media biological activated carbon process for drinking water. *Korean J. Chem. Eng.* **2007**, *24*, 253–260. [CrossRef]
46. Urfer, D.; Huck, P.M. Measurement of biomass activity in drinking water biofilters using a respirometric method. *Water Res.* **2001**, *35*, 1469–1477. [CrossRef]
47. Pharand, L.; Van Dyke, M.I.; Anderson, W.B.; Huck, P.M. Assessment of biomass in drinking water biofilters by adenosine triphosphate. *J. Am. Water Works Assoc.* **2014**, *106*, 433–444. [CrossRef]
48. Niemi, R.M.; Heiskanen, I.; Heine, R.; Rapala, J. Previously uncultured β-proteobacteria dominate in biologically active granular activated carbon (BAC) filters. *Water Res.* **2009**, *43*, 5075–5086. [CrossRef] [PubMed]
49. Persson, F.; Heinicke, G.; Uhl, W.; Hedberg, T.; Hermansson, M. Performance of direct biofiltration of surface water for reduction of biodegradable oganic matter and biofilm formation potential. *Environ. Technol.* **2006**, *27*, 1037–1045. [CrossRef] [PubMed]
50. Velten, S.; Boller, M.; Köster, O.; Helbing, J.; Weilenmann, H.-U.; Hammes, F. Development of biomass in a drinking water granular active carbon (GAC) filter. *Water Res.* **2011**, *45*, 6347–6354. [CrossRef] [PubMed]
51. Xiang, H.; Lu, X.; Yin, L.; Yang, F.; Zhu, G.; Liu, W. Microbial community characterization, activity analysis and purifying efficiency in a biofilter process. *J. Environ. Sci.* **2013**, *25*, 677–687. [CrossRef]
52. Terry, L.G.; Summers, R.S. Biodegradable organic matter and rapid-rate biofilter performance: A review. *Water Res.* **2018**, *128*, 234–245. [CrossRef] [PubMed]
53. Boon, N.; Pycke, B.F.G.; Marzorati, M.; Hammes, F. Nutrient gradients in a granular activated carbon biofilter drives bacterial community organization and dynamics. *Water Res.* **2011**, *45*, 6355–6361. [CrossRef] [PubMed]
54. Liu, X.; Huck, P.M.; Slawson, R.M. Factors Affecting Drinking Water Biofiltration. *J. Am. Water Works Assoc.* **2001**, *93*, 90–101. [CrossRef]
55. Zamorska, J. Biological stability of water after the biofiltration process. *J. Ecol. Eng.* **2018**, *19*, 234–239. [CrossRef]
56. Kolaski, P.; Wysocka, A.; Lasocka-Gomuła, I.; Pruss, A.; Michałkiewicz, M.; Cybulski, Z. Removal of organic matter from water during rapid filtration through a biologically active carbon filter beds—A full scale technological investigation. *Technol. Wody* **2018**, *5*, 8–15. (In Polish)

© 2018 by the authors. Licensee MDPI, Basel, Switzerland. This article is an open access article distributed under the terms and conditions of the Creative Commons Attribution (CC BY) license (http://creativecommons.org/licenses/by/4.0/).

Article

Electroadsorption of Bromide from Natural Water in Granular Activated Carbon

David Ribes [1], Emilia Morallón [2], Diego Cazorla-Amorós [2], Francisco Osorio [3] and María J. García-Ruiz [3,*]

[1] Alicante Water Works, Alona Street 31, 03009 Alicante, Spain; david.ribes@aguasdealicante.es
[2] Institute of Materials, University of Alicante, Apartado 99, 03080 Alicante, Spain; morallon@ua.es (E.M.); cazorla@ua.es (D.C.-A.)
[3] Institute of Water, Department of Civil Engineering, University of Granada, Ramón y Cajal Street, 4, 18071 Granada, Spain; fosorio@ugr.es
* Correspondence: mjgruiz@ugr.es; Tel.: +34-958249463

Abstract: The adsorption and electroadsorption of bromide from natural water has been studied in a filter-press electrochemical cell using a commercial granular activated carbon as the adsorbent. During electroadsorption experiments, different voltages were applied (2 V, 3 V and 4 V) under anodic conditions. The presence of the electric field improves the adsorption capacity of the activated carbon. The decrease in bromide concentration observed at high potentials (3 V or 4 V) may be due to the electrochemical transformation of bromide to Br_2. The anodic treatment produces a higher decrease in the concentration of bromide in the case of cathodic electroadsorption. Moreover, in this anodic electroadsorption, if the system is again put under open circuit conditions, no desorption of the bromide is produced. In the case of anodic treatment in the following adsorption process after 24 h of treatment at 3 V, a new decrease in the bromide concentration is observed as a consequence of the decrease in bromide concentration after the electrochemical stage. It can be concluded that the electroadsorption process is effective against the elimination of bromide and total bromine in water, with a content of 345 and 470 $\mu g\ L^{-1}$, respectively, reaching elimination values of 46% in a single-stage electroadsorption process in bromide and total bromine. The application of the electric field to the activated carbon with a positive polarization (anodic electroadsorption) increases the adsorption capacity of the activated carbon significantly, achieving a reduction of up to 220 $\mu g\ L^{-1}$ after 1 h of contact with water. The two stage process in which a previous electrochemical oxidation is incorporated before the electroadsorption stage significantly increased the efficiency from 46% in a single electroadsorption step at 3 V, to 59% in two stages.

Keywords: electroadsortion; granular activated carbon; thrialomethanes; bromides

1. Introduction

Disinfection and elimination of pathogens is a fundamental part of water treatment for human consumption, with chlorination being the most widespread method in the world due to its low cost, among other favorable conditions. To ensure water quality, a concentration between 0.2 and 1 mg L^{-1} of free residual chlorine is necessary in the entire drinking water distribution network. When chlorine is combined with some contaminants present in the water, disinfection by-products can be generated. These water disinfection by-products (DBPs) are substances that are formed as a result of the reaction between disinfectants and some compounds present in water, such as natural organic matter. Organic matter dissolved in raw water is one of the main causes of the formation of DPBs, since it reacts with chlorine giving rise to the formation of halogenated trihalomethanes [1]. Trihalomethanes are disinfection by-products (DBPs) formed by the reaction of chlorine derivatives in drinking water with their precursors, which can be organic matter or bromide. The most common THMs are chloroform trichloromethane ($CHCl_3$),

Citation: Ribes, D.; Morallón, E.; Cazorla-Ruiz, D.; Osorio, F.; García-Ruiz, M.J. Electroadsorption of Bromide from Natural Water in Granular Activated Carbon. *Water* **2021**, *13*, 598. https://doi.org/10.3390/w13050598

Academic Editor: Chicgoua Noubactep

Received: 20 December 2020
Accepted: 16 February 2021
Published: 25 February 2021

Publisher's Note: MDPI stays neutral with regard to jurisdictional claims in published maps and institutional affiliations.

Copyright: © 2021 by the authors. Licensee MDPI, Basel, Switzerland. This article is an open access article distributed under the terms and conditions of the Creative Commons Attribution (CC BY) license (https://creativecommons.org/licenses/by/4.0/).

 check for updates

bromodichloromethane (CHBrCl$_2$), dibromochloromethane (CHBr$_2$Cl), bromoform, and tribromomethane (CHBr$_3$) [2].

Environmental legislation regarding the presence of trihalomethanes (THMs) in drinking water is becoming increasingly restrictive. In accordance with current regulations, the need to control the concentration of these compounds in water for human consumption is established, since their presence can cause serious damage to human health [3].

In this sense, drinking water, or sources of water intended for human consumption (potable and/or pre-potable), usually have a low concentration of bromide. However, sometimes they can be generated during the purification process and lead to compounds of high toxicity for humans [4–7].

The presence of bromide is of special interest because, unlike chloride, bromide favors the formation of THMs, producing compounds such as bromoform (CHBr$_3$), bromodichloromethane (CHCl$_2$Br) or chlorodibromomethane (CHClBr$_2$), among others. [8,9]. Bromide is found naturally in seawater and in coastal areas, which is a problem of special interest in the waters for human consumption in the Mediterranean basin.

Conventional water treatment processes (coagulation, flocculation, and sedimentation) remove organic precursors from water in the range of 20%–60% [10–12], but they are not effective in removing inorganic precursors [13–17]. Bromide, arsenic, phosphate, nitrite and selenium are available in water as anions. Therefore, the removal is strongly dependent on the hydrophobic character of the sorbent (in essence it is about the isolectric point and/or point of zero charge) and the properties of the anions (in this sense, the speciation that encompasses the charge and the changes are important with respect to electrostatic interaction with the adsorbent). Thus, adsorption is favored in hydrophobic materials, such as activated carbons, for hydrated anions with a low degree of hydration, when comparing anions with the same charge [18,19]. In this study, we focus on the removal of bromide in waters with elevated bromide levels. This results in an increase in the Br-/DOC ratio and promotes the formation of brominated DBPs with higher toxicity [20]. Thus, new and advanced treatment technologies are needed to remove inorganic precursors.

Until now, the existing treatment systems for the removal of Br$^-$, I$^-$, and Cl$^-$ ions are based on the use of different types of ion exchange resins, although they are not very efficient due to their high cost and the low selectivity of the exchange process [21–25]. Activated carbons are the alternative for the removal of organic and inorganic pollutants from polluted waters due to the combination of high surface area, chemical and physical stability, adjustable surface chemistry and reduced production costs. Thus, adsorption treatments using these porous carbon materials are known to combine a high degree of efficiency with a reasonable cost [26,27].

GAC has a number of properties, such as high surface area, large porosity, well developed internal pore structure consisting of micro-, meso- and macropores, as well as a wide spectrum of functional groups present on the surface of AC, which makes it a versatile material that has numerous applications in many areas, but mainly in the environmental field [28]. The efficiency of ACs as adsorbents for diverse types of pollutants is well reported [29,30]. It is well known that activated carbon has been found to be very efficient in removing organic compounds than metals and other inorganic pollutants. Efforts are ongoing to substantially improve the potential of carbon surface by using different chemicals or suitable treatment methods [31], which will enable AC to enhance its potential for the removal of specific contaminants from the aqueous phase. The use of granular activated carbons in organic DBP precursors removal has been widely reported. [32] However, the influence of carbon surface tailoring on Br removal has not been extensively investigated, and very few studies have explored Br removal by the tailored AC surface at practical adsorbent doses [33], along with the control of brominated DBPs in natural waters.

In this sense, activated carbon treatments with silver ions deposited on their surface have been used for selective adsorption of Br$^-$, I$^-$, and Cl$^-$ anions, although their adsorption efficiency is not very important due to the dissolution of silver during the process [34,35]. The use of granular activated carbon (GAC) adsorption is beneficial from a standpoint of controlling disinfection by-product (DBP) formation and associated toxicity in water impacted by bromide and iodide. GAC has been used as an effective material to control DBP formation on a large range of bromide (20–1000 $\mu g\ L^{-1}$) and iodide (<5–100 $\mu g\ L^{-1}$) concentrations [36]. There are many reports in the literature on the use of carbon adsorbents for the removal of metals and metal compounds, but generally not for low $\mu g\ L^{-1}$ levels [34,35,37], which is mainly a consequence of its hydrophobic nature. Moreover, composite materials based on activated carbons have been used in the removal of inorganic compounds in water [18,19,38].

More particularly, remediation of metallic ions in concentrations in the range of parts per million (mg L^{-1}) using activated carbon as the adsorbent has proved to be feasible [39–41]. Interestingly, under certain conditions, the effectiveness of activated carbons can be significantly improved by the application of an electric field during the adsorption/desorption process (electrosorption/electrodesorption), which may enhance their capacity [42,43]. In those cases, the application of a controlled electric potential to the adsorbent can enhance adsorption by enabling the formation of an electric double layer in the surface of the activated carbon in contact with the solution. This phenomenon shares the same fundamentals as the energy storage in supercapacitors [44] and the capacitive deionization for water purification [42,45], and it is only possible in porous materials with adequate pore size distribution and acceptable conductivity, such as activated carbons. However, most of the electroadsorption studies have been done for organic compounds [43,46,47], although it has also been used in the elimination of inorganic salts in water [48,49]; however, this electroadsorption methodology has not been used for the elimination of bromide in water in a low concentration level of $\mu g\ L^{-1}$. The electroadsorption process, under appropriate conditions, may be effective for the removal of a low concentration of pollutants, and in addition, it is a non-destructive procedure.

This research studies the removal of Br^- ion and total bromine in water by electroadsorption and the combination of this process with an electrochemical treatment under anodic conditions using an activated carbon filter in an electrochemical cell, thus suppressing the possible formation of these brominated compounds. In this sense, the main objective of this work is to study the reduction in the concentration of bromide and total bromine in a solution that allows, in subsequent investigations, it to develop/optimize a procedure for the elimination of bromide to comply with current regulations.

2. Materials and Methods

2.1. Characterization of Granular Activated Carbon

In this study, the activated carbon used was a Granular Activated Carbon (GAC), available from Waterlink Suctliffe Carbons (Lancashire, UK). It is a high porosity material whose adsorption was characterized using the Autosorb-6 equipment (Quantachrome, Boynton Beach, FL, USA). For this, the adsorption-desorption analyses of N_2 were performed at a temperature of $-196\ ^\circ C$ and CO_2 adsorption at a temperature of $0\ ^\circ C$. Previously, the sample was degassed under a vacuum for 4 h at a temperature of $250\ ^\circ C$. To determine the apparent surface area S_{BET}, the BET equation was applied to the adsorption isotherm of N_2 under the relative pressures region between 0.05 and 0.20. Similarly, for the determination of the total volume of micropores, the Dubinin Radushkevich equation (DR) was applied on the adsorption isotherm of N_2 in the range of relative pressures between 0.005 and 0.10. Regarding the narrow volume of micropores, it was determined by means of the CO_2 adsorption isotherm [50]. Table 1 shows the textural properties of the GAC.

Table 1. Characterization of the porosity of Activated Carbon PQ-0602-04.

S_{BET} (m^2/g)	V_{N2} (cm^3/g)	V_{CO2} (cm^3/g)
865	0.36	0.25

Granular activated carbon was thoroughly washed using distilled water, filtered and dried at 110 °C in order to clean its internal surface before its characterization and further use as bromide adsorbent.

2.2. Water Characterization, Bromide and Total Bromine Determination

The drinking water used in this study is a real sample that came from the Alicante seawater desalination plant (Alicante, Spain). The bromide concentration (Br^-) was 345 µg L^{-1}, and this concentration was analyzed by ion chromatography. The Br^- and total bromine (Br) concentrations during the different experiments were measured using inductively coupled plasma-mass spectrometry (ICP-MS) with Perkin-Elmer 4300 DV equipment (Waltham, MA, USA). For the ICP-MS measurements, each sample was dissolved in 0.5 M HNO_3, followed by filtration using a nylon membrane filter (pore diameter ~350 nm). The total bromine concentration includes all Br-containing species present in the water.

The water characterization is detailed in the following table (Table 2).

Table 2. Water characterization.

Parameters	Value	Units
Bromine	470	µg L^{-1}
Total Organic Carbon	<0.5	mg L^{-1}
Conductivity (20 °C)	562	µS cm^{-1}
pH	8.6	U. pH
Bromate	25	µg L^{-1}
Bromide	345	µg L^{-1}

2.3. Electrochemical Filter-Press Cell

In this study, the filter press electrochemical cell for the electroadsorption of bromide has been developed. The electrochemical cell has been modified in order to put it in contact with the contaminated water with a wider activated carbon bed [51]. During the experiment, the water was circulated by means of a centrifugal pump between the anode and the cathode and passing through the activated carbon that is found as a bed. In this way, information can be obtained on the adsorption kinetics in this batch reactor, as well as the amount adsorbed at the equilibrium. This last data is very important to know the possible improvement in the adsorption capacity of the activated carbon as a consequence of the applied potential. Figure 1 shows a diagram of the electrochemical cell and the device used for the electroadsorption process.

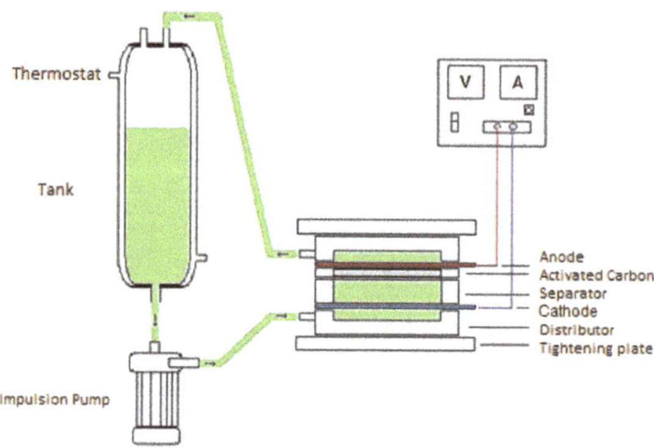

Figure 1. Schematic diagram of the electroadsorption experiments in which the electrochemical chemical cell together with the current source, water tank and impulsion pump are included.

It is divided into two compartments that are herein referred to as the anolyte and catholyte, and each one being composed by (i) a flow distributor with one nozzle that can be used as a water inlet or outlet, (ii) a mesh for keeping the activated carbon particles and the electrode pressed altogether, (iii) a joint that provides volume for allocating the activated carbon particles, and (iv) an electrode. The compartments are tightly connected together by means of (v) two screwed stainless steel plates, with (vi) a final joint being displayed between anode and cathode. In this work, the electrochemical filter-press cell was disposed horizontally without compartment separation (i.e., no selective ionic membrane was used between the compartments). The cell has a plane electrode area of 20 cm^2. The width of each compartment is 0.9 cm, resulting in a volume of 18 cm^3. The external DC Power supply was Blausonic FA-325 (Promax Test & Measurement, Barcelona, Spain), 30 V, and 2.5 A.

2.4. Operational Conditions

In this study, the following operational conditions were carried out with 4 g of the washed GAC placed near the cathode or near the anode for the cathodic or anodic electroadsorptions, respectively, and contacted with 400 mL of water [52]. While the contact between the GAC particles and the electrode was slightly loose, in previous research it was proved that, for a configuration similar to that used in this study, a polarization of carbon particles is achieved [53]. The adsorption experiments were initiated when real Br-containing water started circulating inside the electrochemical cell for 24 h in the absence of current, which was enough time to ensure the adsorption equilibrium (around 250 $\mu g\ L^{-1}$). In order to analyze the effect of the voltage, different values have been used in different experiments, with new water, during 24 h in order to reach the equilibrium conditions at each voltage. Anodic cathodic conditions in other experiments, after the adsorption in the absence of voltage, at three different voltages, 2.0, 3.0, and 4.0 V, were sequentially applied. Each voltage step was imposed for 24 h before proceeding to the next electroadsorption stage in order to reach the equilibrium conditions. In this case, no additional water is used for electroadsorption in order to achieve the simplest possible bromide removal treatment. In order to analyze the reversibility of Br^- removal in the electroadsorption experiments, one last step was carried out in the absence of voltage during 24 h. The analysis of the bromide and bromine were determined, as exposed above, and 1 mL of solution was periodically taken from the electrochemical cell.

3. Results

The selection of this material, GAC, and the real natural water sample used will allow us to demonstrate the validity of this method at conditions very close to those available in a real water treatment plant. The most relevant surface properties of the GAC were assessed by N_2 and CO_2 adsorption–desorption isotherms. In this sense, the micropore volume obtained from CO_2 adsorption, compared to the N_2 one, indicates that most of the microporosity has a size of about 0.5 nm (the micropore volumes from N_2 and CO_2 adsorptions are similar (see Table 1) [50]. Thus, from the point of view of textural properties, it can be concluded that the selected GAC has a well-developed porous texture, which makes it a good choice for water treatment.

3.1. Effect of Voltage in the Anodic Electroadsorption

In order to initially identify the amount adsorbed by the activated carbon in the absence of voltage, and thus check if the presence of electric field produces an increase in the amount adsorbed, the water (400 mL) was put in contact in the filter-press cell with activated carbon (4 g) and allowed to reach the adsorption equilibrium for 24 h Figure 2 shows the variation of the concentration with time; in this figure, the initial concentration of bromide and total bromine are also represented.

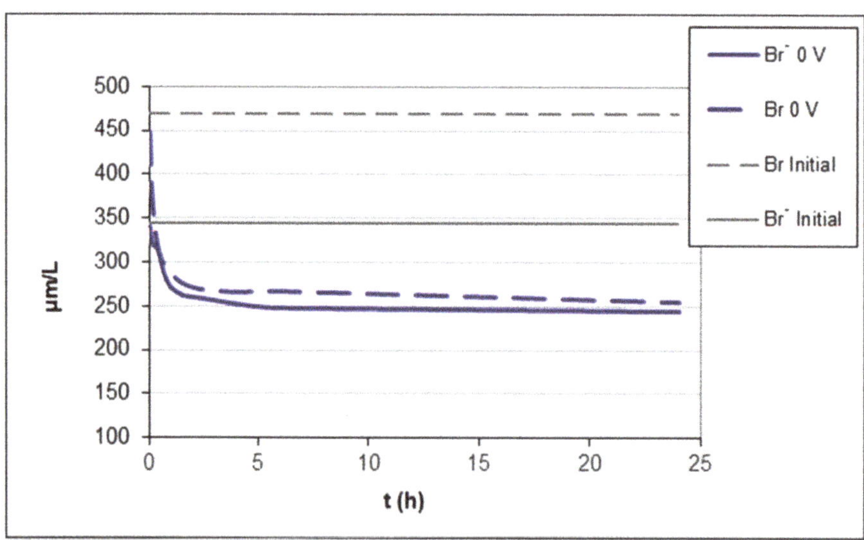

Figure 2. Variation of bromide and total bromine concentration in the absence of voltage. Filter press.

It is observed that the concentration of bromide and total bromine decreases as a consequence of adsorption on activated carbon until reaching an equilibrium concentration of approximately 250 μg L^{-1} for the bromide. Figure 2 also shows the presence of two regions in the adsorption process. In the first region, the adsorption is very fast, and it is mainly due to either mass transfer through the stationary liquid boundary layer that appears over the external surface of the particle or restricted pore diffusion of the adsorbate in the wider porosity of the carbon particle. A second region is distinguished after 7 h, where adsorption is slower due to surface diffusion of adsorbed molecules within narrower pores. The appearance of two zones on the adsorption graph of an activated carbon is very characteristic of these materials [54,55].

The initial Br- concentration is 345 μg L^{-1}, and the equilibrium concentration reached is 250 μg L^{-1}. Therefore, the adsorption capacity of the activated carbon in these conditions is 9.5 μg g^{-1}, which corresponds to a removal efficiency of 29% (Table 3).

Table 3. Bromide concentration at the end of the experiment, % removal and amount adsorbed at the equilibrium on activated carbon in different electroadsorption experiments at different voltages.

Voltage (V)	[Br-] Final (μg L^{-1})	% Removal	q (μg/g)
Open Circuit	244	29	9.5
2 V	205	41	14.0
3 V	186	46	15.9
4 V	202	41	14.4

Figure 3 shows the bromide and total bromine concentration during electroadsorption when different voltages are applied (2 V, 3 V and 4 V) under anodic conditions. It can be observed that the concentrations of bromide and total bromine decrease with time for all voltages used and lower values are reached than those obtained in the absence of the voltage (0 V in Figure 3). These results indicate that the presence of the electric field increases the amount of bromide and total bromine removal, improving the adsorption capacity of the activated carbon.

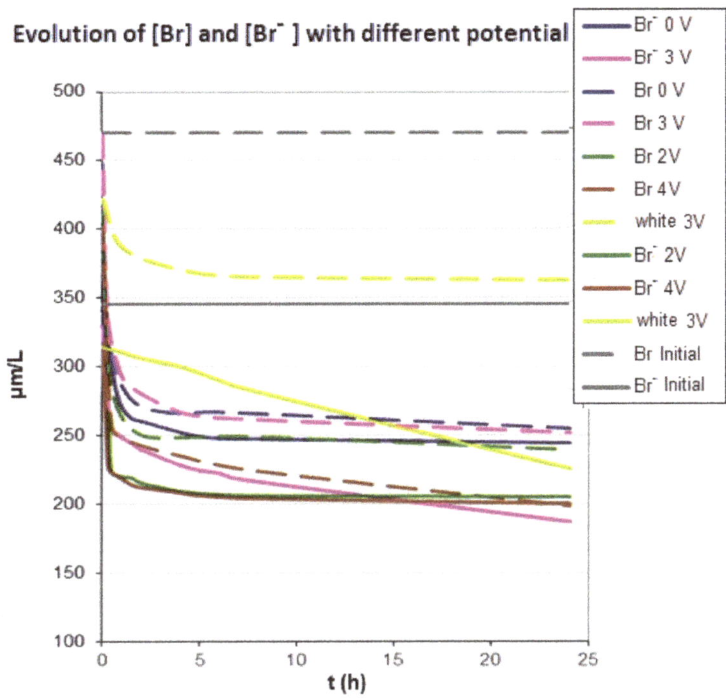

Figure 3. Variation of the bromide and total bromine at different voltages (2 V, 3 V and 4 V) under anodic conditions. The adsorption in the absence of voltage is also included. The dashed line corresponds to the total bromine and full line to the bromide.

We can also observe that, in the case of 4 V, the bromide and total bromine concentrations decrease significantly and with a much higher rate than in the absence of potential, achieving a reduction of the bromide concentration by 29% after 0.5 h of contact with the activated carbon. In the absence of voltage, the reduction in concentration after 0.5 h is 14 %. The bromide concentration after equilibrium is 200 µg L^{-1}, which corresponds to an adsorption capacity of the activated carbon of 14.5 µg g^{-1} and to a reduction of the bromide concentration of 42%.

At high potentials, bromide ions can be oxidized to Br_2 according to Equation (1), and the decrease in concentration observed at high potentials (3 V or 4 V) may be due to the electrochemical transformation of bromide to bromine. At these voltages, values higher than the standard potential of this reaction are reached in the anode, making the oxidation of bromide possible. It should be noted that the adsorption of Br_2 on activated carbon is a very favorable process, unlike the anion.

$$Br_2 + 2e^- \rightarrow 2\,Br^-\quad E^0 = 1.08V/ENH \tag{1}$$

To verify this result, a pure electrochemical treatment experiment in the absence of activated carbon in the electrochemical cell was carried out by applying a voltage of 3 V, and the result is shown in Figure 4. This voltage is selected in order to avoid the formation of bubbles on the electrode.

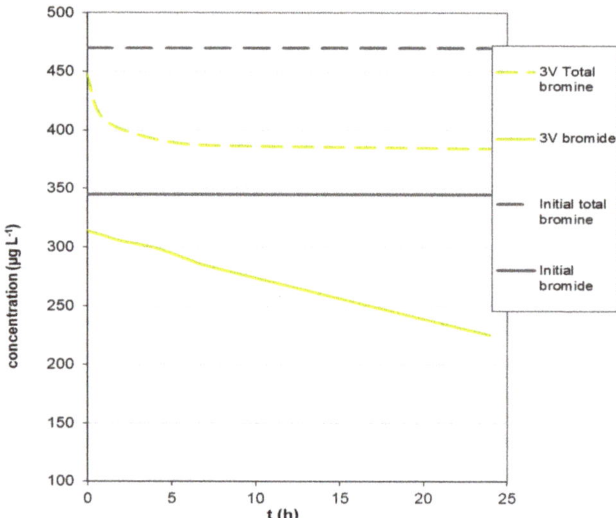

Figure 4. Variation of bromide concentration in an experiment in the absence of activated carbon at 3 V.

It can be observed, in Figure 4, that the bromide concentration decreases continuously over time, indicating that it is oxidized on the electrode, thus decreasing its concentration in the solution; however, the total bromine after an initial decrease, probably due to some volatilization, remains constant with time, indicating the conversion of bromide into Br_2 according to Equation (1). This oxidation reaction, once the constant value of bromine is reached, is not producing a change in the total bromine in solution, but an interconversion of one species into another.

Tables 3 and 4 compare the final bromide and final total bromine concentration values obtained in the adsorption and electroadsorption experiments after 24 h at different voltages. The tables also include the percentages of bromide removal and total bromine removal (with respect to the initial concentration). In addition, the amount adsorbed by the activated carbon of both bromide and total bromine is shown.

Table 4. Total bromine concentration at the end of the experiment, % removal and amount adsorbed at equilibrium on activated carbon in different electroadsorption experiments.

Voltage (V)	[Br] Final (μg L^{-1})	% Removal	q (μg/g)
Open Circuit	255	46	20.0
2 V	248	47	20.7
3 V	252	46	20.3
4 V	198	58	25.7

We can see that the bromide concentration decreases in the absence of potential by 29% [33] and we managed to increase the percentage of elimination by 41% and 46% in the presence of the electric field, which is 12% and 17% more than just in a process of adsorption (open circuit conditions). However, the value of total bromine content in adsorption and electroadsorption is similar (Table 4) to the applied voltage up to 3 V.

However, when the voltage increases to values where the electrode potential reaches the oxidation of bromide to bromine (4 V), the percentage of total bromine removal increases, which indicates that the adsorption of bromine is more favorable on the activated carbon increasing the adsorption capacity. However, at this voltage, the formation of gas on the electrode surface is clearly observed, mainly due to water oxidation. Then, in order to avoid the formation of bubbling on the electrode, the effect of polarity on the electroadsorption, a voltage of 3 V has been selected.

Energy Consumption of Electrochemical Removal of Bromide

The Energy Consumption (EC) per unit volume of treated water was calculated according to Equation (2):

$$EC(kWh\ L^{-1}) = (U \cdot I \cdot t)/(V \cdot 1000) \tag{2}$$

where U is the voltage of the electrochemical cell (V), I is the average current (A), t is the treatment time (h) and V is the volume of treated water (L).

Then, the EC during the removal of bromide in this laboratory scale is between 1 and 2 $kWh\ L^{-1}$.

3.2. Effect of the Polarity of the Electrode (Anodic Electroadsorption and Cathodic Electroadsorption)

In order to study the effect of the polarity of the electroadsorption experiments, the comparison between the performance of an anodic electroadsorption (the activated carbon is located in the anode compartment in which the activated carbon is subjected to a positive polarity) and the cathodic electroadsorption (the activated carbon is located in the cathode compartment, in which the activated carbon is subjected to a negative polarity) has been done. In these experiments, the following consecutive steps have been done: (i) adsorption at open circuit conditions during 24 h, (ii) application a voltage of 3 V during 24 h and (iii) again open circuit potential during 24 h in order to analyze the reversibility of the electroadsorption step.

Figure 5 shows the evolution of the bromide concentration during these experiments at cathodic and anodic conditions during the second step. It can be observed that, after the adsorption at the open circuit potential (step 1 in Figure 5), the amount of bromide is similar in both cases; however, after the electroadsorption, the anodic treatment produces a higher decrease in the concentration of bromide that, in the case of cathodic electro-adsorption (step 2 in Figure 5). These results can be as a consequence of two processes: the electroadsorption of the bromide that is favored at a positive polarization of the active carbon and the oxidation of bromide on the anode according to Equation (1). Moreover, after the electroadsorption step, if the system is again put at open circuit conditions, no desorption of the bromide is produced in both polarities, indicating the irreversibility of the electroadsorption processes.

It can be also seen that, after the anodic electroadsorption treatment (24 h of treatment at 3 V), in the following adsorption process at open circuit potential (step 3 in Figure 5), a new decrease in the bromide concentration is observed. These results seem to indicate that the bromide has been oxidized in anodic conditions to Br_2 and a new adsorption equilibrium of bromide is again reached as a consequence of the different concentration of bromide in solution.

3.3. Adsorption after an Electrochemical Oxidation Process: Two-Stages Electroadsorption

In order to improve the bromide removal, the following experiment in two stages was designed: in the first, an electrochemical treatment during 24 h takes place in the absence of activated carbon (electrolysis stage). During this stage, the bromide is oxidized to Br_2, which can be easily adsorbed on the activated carbon. Then, after this electrochemical oxidation stage, the water is passed to another electrochemical cell in which a second adsorption stage during 24 h in the presence of activated carbon is done, but in the absence of potential (adsorption stage). The water is continuously circulated between both electrochemical cells and the concentration is taken over time. Figure 6 shows a scheme of the electrochemical device used.

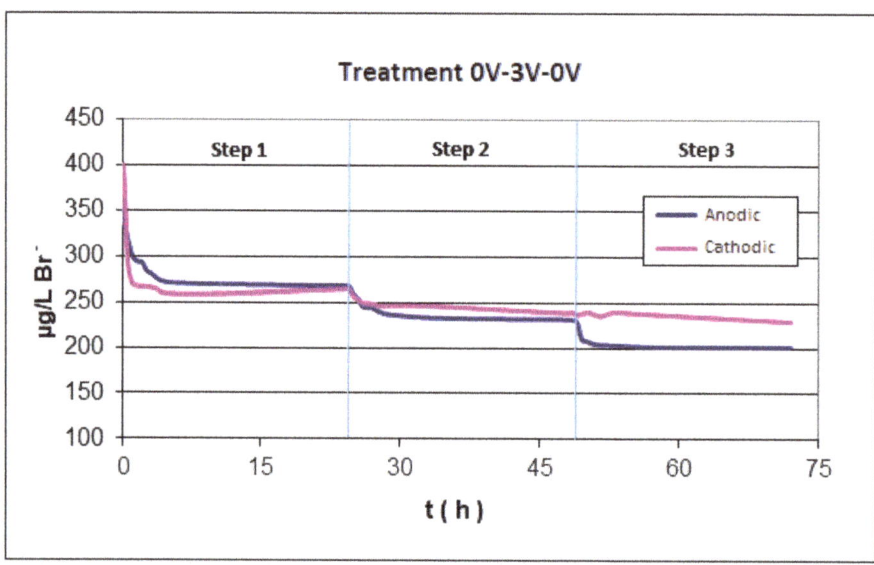

Figure 5. Bromide concentration during three steps experiment, the second corresponds to anodic and cathodic electroadsorption experiments for 24 h at 3 V. 3 V is applied for 24 h in the second step and then it is left at open circuit again for 24 h.

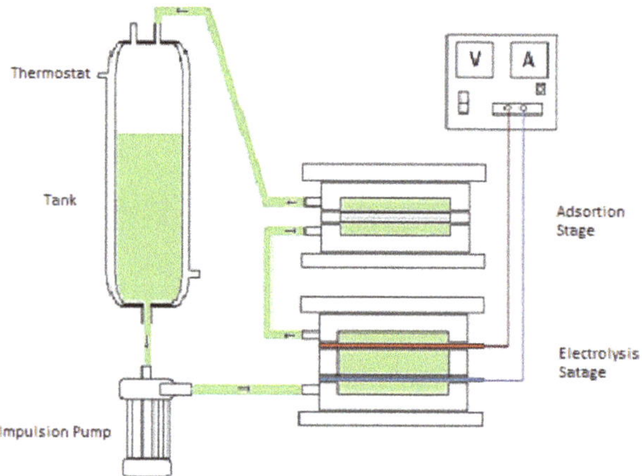

Figure 6. Scheme of the two electroadsorption stages.

The data obtained after the total treatment are shown in Table 5. We can see that the concentration of bromide reaches a value similar to that obtained in the pure electroadsorption experiment in the same conditions of voltage (Table 3); however, the total bromine falls below 200 μg L^{-1} and the percentage of elimination of this total bromine increases with respect to that obtained in an electroadsorption process (Table 4). Then, using two stages in comparison to only the electroadsorption stage, it is possible to reduce both concentrations, bromide and total bromine, from the water.

Table 5. Bromide concentration, bromine and percent removal and amount adsorbed by activated carbon in a two-stage experiment at the end of the treatment.

	[Br-] Final (μg L^{-1})	% Removal	q (μg/g)
2 Stages	185	46	16.0
	[Br] Final(μg L^{-1})	% Removal	q (μg/g)
2 Stages	194	59	25.8

4. Conclusions

It can be concluded that the electroadsorption process is effective against the elimination of bromide and total bromine in water with a content of 345 μg L^{-1} and 470 μg L^{-1}, respectively, reaching elimination values of around 46 % in a single-stage electroadsorption process in both bromide and total bromine. It has been studied that the better conditions for removal of bromide in water with a concentration is the application of the electric field to the activated carbon that produces a positive polarization (anodic electroadsorption). This treatment increases the adsorption capacity of the activated carbon significantly, achieving a reduction of bromide of up to 220 μg L^{-1} after 1 h of contact with water. The removal percentage of bromide increases from 29% in the absence of voltage (adsorption) to 46% at 3 V (electroadsorption). When the process is carried out in two stages in which a previous electrochemical oxidation to the adsorption stage is introduced, similar values of removal of bromide are achieved. In this previous electrochemical oxidation stage, the oxidation of bromide to Br$_2$, which is easily adsorbed on the activated carbon, is produced, and the total bromine content of the water is reduced by 59% in the two stages of the experiment, with this value significantly increased from 46% in a single electroadsorption step at 3 V.

Author Contributions: D.R.: Investigation, Writing—Original Draft; E.M.: Conceptualization, Methodology, Writing—Review & Editing; D.C.-A.: Investigation, Writing—Original Draft; Data Curation; F.O.: Conceptualization, Writing—Review & Editing, Supervision; M.J.G.-R.: Conceptualization, Writing—Review & Editing, Supervision. All authors have read and agreed to the published version of the manuscript.

Funding: This research received no external funding.

Institutional Review Board Statement: Not applicable.

Informed Consent Statement: Not applicable.

Data Availability Statement: Not applicable.

Conflicts of Interest: The authors declare no conflict of interest.

References

1. Gopal, K.; Tripathy, S.S.; Bersillon, J.L.; Dubey, S.P. Chlorination byproducts, their toxicodynamics and removal from drinking water. *J. Hazard. Mater.* **2007**, *140*, 1–6. [CrossRef]
2. Symons, J.M.; Stevens, A.; Clark, R.M.; Geldreich, E.E.; Love, O.T., Jr.; DeMarco, J. *Treatment Techniques for Controlling Trihalomethanes in Drinking Water*; EPA/600/2-81/156 (NTIS PB82163197); U.S. Environmental Protection Agency: Washington, DC, USA, 2002.
3. Villanueva, C.M.; Cordier, S.; Font-Ribera, L.; Salas, L.A.; Levallois, P. Overview of disinfection by-products and associated health effects. *Curr. Environ. Health Rep.* **2015**, *2*, 107–115. [CrossRef]
4. Cortés, C.; Marcos, R. Genotoxity of disinfection byproducts and disinfected waters: A review of recent literature. *Mutat. Res./Toxicol. Environ. Mutagenesis* **2018**, *831*, 1–12. [CrossRef] [PubMed]
5. Liviac, D.; Wagner, E.D.; Mitch, W.A.; Altonji, M.J.; Plewa, M.J. Genotoxicity of Water Concentrates from Recreational Pools after Various Disinfection Methods. *Environ. Sci. Technol.* **2010**, *44*, 3527–3532. [CrossRef]
6. Plewa, M.J.; Wagner, E.D.; Richardson, S.D.; Thruston, A.D.; Woo, Y.-T.; McKague, A.B. Chemical and biological char-acterization of newly discovered iodoacid drinking water disinfection byproducts. *Environ. Sci. Technol.* **2004**, *38*, 4713–4722. [CrossRef] [PubMed]

7.   Plewa, M.; Wagner, E. Risks of Disinfection Byproducts in Drinking Water: Comparative Mammalian Cell Cytotoxicity and Genotoxicity. In *Encyclopedia of Environmental Health*; Elsevier: Amsterdam, The Netherlands, 2019; Volume 5, pp. 559–566. [CrossRef]

8.   Beita-Sandí, W.; Selbes, M.; Kim, D.; Karanfil, T. Removal of N-nitrosodimethylamine precursors by cation exchange resin: The effects of pH and calcium. *Chemosphere* **2018**, *211*, 1091–1097. [CrossRef]

9.   Soyluoglu, M.; Ersan, M.S.; Ateia, M.; Karanfil, T. Removal of bromide from natural waters: Bromide-selective vs. conventional ion exchange resins. *Chemosphere* **2020**, *238*, 124583. [CrossRef]

10.   Joseph, L.; Flora, J.R.; Park, Y.-G.; Badawy, M.; Saleh, H.; Yoon, Y. Removal of natural organic matter from potential drinking water sources by combined coagulation and adsorption using carbon nanomaterials. *Sep. Purif. Technol.* **2012**, *95*, 64–72. [CrossRef]

11.   Zainudin, F.M.; Abu Hasan, H.; Abdullah, S.R.S. An overview of the technology used to remove trihalomethane (THM), trihalomethane precursors, and trihalomethane formation potential (THMFP) from water and wastewater. *J. Ind. Eng. Chem.* **2018**, *57*, 1–14. [CrossRef]

12.   Sillanpää, M.; Ncibi, M.C.; Matilainen, A.; Vepsäläinen, M. Removal of natural organic matter in drinking water treatment by coagulation: A comprehensive review. *Chemosphere* **2018**, *190*, 54–71.

13.   Ates, N.; Yilmaz, L.; Kitis, M.; Yetis, U. Removal of disinfection by-product precursors by UF and NF membranes in low-SUVA waters. *J. Membr. Sci.* **2009**, *328*, 104–112. [CrossRef]

14.   Kristiana, I.; Joll, C.; Heitz, A. Powdered activated carbon coupled with enhanced coagulation for natural organic matter removal and disinfection by-product control: Application in a Western Australian water treatment plant. *Chemosphere* **2011**, *83*, 661–667. [CrossRef]

15.   Singer, P.C.; Bilyk, K. Enhanced coagulation using a magnetic ion exchange resin. *Water Res.* **2002**, *36*, 4009–4022. [CrossRef]

16.   Xu, Z.; Jiao, R.; Liu, H.; Wang, D.; Chow, C.W.K.; Drikas, M. Hybrid treatment process of using MIEX and high per-formance composite coagulant for DOM and bromide removal. *J. Environ. Eng.* **2013**, *139*, 79–85. [CrossRef]

17.   Yang, S.; He, M.; Zhi, Y.; Chang, S.X.; Gu, B.; Liu, X.; Xu, J. An integrated analysis on source-exposure risk of heavy metals in agricultural soils near intense electronic waste recycling activities. *Environ. Int.* **2019**, *133*, 105239. [CrossRef] [PubMed]

18.   Rabiul, M. Efficient phosphate removal from water for controlling eutrophication using novel composite adsorbent. *J. Clean. Prod.* **2019**, *228*, 1311–1319.

19.   Rabiul, M.; Munjur, M.; Asiri, A.; Rahma, M. Cleaning the arsenic (V) contaminated water for safe-guarding the public health using novel composite material. *Compos. Part B Eng.* **2019**, *171*, 294–301.

20.   Phetrak, A.; Lohwacharin, J.; Sakai, H.; Murakami, M.; Oguma, K.; Takizawa, S. Simultaneous removal of dissolved organic matter and bromide from drinking water source by anion exchange resins for controlling disinfection by-products. *J. Environ. Sci.* **2014**, *26*, 1294–1300. [CrossRef]

21.   Boyer, T.H.; Singer, P.C. Bench-scale testing of a magnetic ion exchange resin for removal of disinfection by-product precursors. *Water Res.* **2005**, *39*, 1265–1276. [CrossRef] [PubMed]

22.   Hsu, S.; Singer, P.C. Removal of bromide and natural organic matter by anion exchange. *Water Res.* **2010**, *44*, 2133–2140. [CrossRef]

23.   Neale, P.; Schäfer, A. Magnetic ion exchange: Is there potential for international development? *Desalination* **2009**, *248*, 160–168. [CrossRef]

24.   Walker, K.M.; Boyer, T.H. Long-term performance of bicarbonate-form anion exchange: Removal of dissolved organic matter and bromide from the St. Johns River, FL, USA. *Water Res.* **2011**, *45*, 2875–2886. [CrossRef]

25.   Rabiul, M.; Jyo, A.; El-Safty, S.; Tamada, M.; Seko, N. A weak-base fibrous anion exchanger effective for rapid phosphate removal from water. *J. Hazard. Mater.* **2011**, *188*, 164–171.

26.   Derbyshire, F.; Jagtoyen, M.; Andrews, R.; Rao, A.; Martín-Gullón, I.; Grulke, E. Carbon materials in environmental applications In *Chemistry and Physics of Carbon*; Radovic, L.R., Ed.; Marcel Dekker: New York, NY, USA, 2001; Volume 27, pp. 1–66.

27.   Radovic, L.R.; Moreno-Castilla, C.; Rivero Utrilla, J. Carbon materials as adsorbents in aqueous solutions. *Chem. Phys. Carbon* **2000**, *27*, 227–405.

28.   Bhatnagar, A.; Hogland, W.; Marques, M.; Sillanpää, M. An overview of the modification methods of activated carbon for its water treatment applications. *Chem. Eng. J.* **2013**, *219*, 499–511. [CrossRef]

29.   Hassler, J.W.; Cheremisinoff, P.N. *Carbon Adsorption Handbook*; Ann Arbor Science: Ann Arbor, MI, USA, 1980.

30.   Yang, R.T. *Adsorbents: Fundamentals and Applications*; John Wiley & Sons: Hoboken, NJ, USA, 2003.

31.   Monser, L.; Adhoum, N. Modified activated carbon for the removal of copper, zinc, chromium and cyanide from wastewater. *Sep. Purif. Technol.* **2002**, *26*, 137–146. [CrossRef]

32.   Owen, D.M. *Removal of DBP Precursors by GAC Adsorption*; American Water Works Association: Denver, CO, USA, 1998.

33.   Watson, K.; Farré, M.J.; Knight, N. Comparing a silver-impregnated activated carbon with an unmodified activated carbon for disinfection by-product minimisation and precursor removal. *Sci. Total. Environ.* **2016**, *542*, 672–684. [CrossRef]

34.   Pattanayak, J.; Mondal, K.; Mathew, S.; Lalvani, S.B. A parametric eval-cuation of the removal of As(V) and As(III) by carbon based adsorbents. *Carbon* **2000**, *38*, 589–596. [CrossRef]

35.   Babic, B.M.; Milonjic, S.K.; Polovina, M.J.; Cupic, S.; Kaludjerovic, B.V. Adsorption of zinc, cadmium and mercury ions from aqueous solution sonan activated carbon cloth. *Carbon* **2002**, *40*, 1109–1115. [CrossRef]

36. Zhang, C.; Maness, J.C.; Cuthbertson, A.A.; Kimura, S.Y.; Liberatore, H.K.; Richardson, S.D.; Stanford, B.D.; Sun, M.; Knappe, D.R. Treating water containing elevated bromide and iodide levels with granular activated carbon and free chlorine: Impacts on disinfection byproduct formation and calculated toxicity†. *Environ. Sci. Water Res. Technol.* **2020**, *6*, 3460. [CrossRef]

37. Namasivayam, C.; Kadirvelu, K. Uptake of mercury (II) from wastewater by activated carbon from an unwanted agri-cultural solid by-product: Coirpith. *Carbon* **1999**, *37*, 79–84. [CrossRef]

38. Rabiul, M. A novel facial composite adsorbent for enhanced copper(II) detection and removal from wastewater. *Chem. Eng. J.* **2015**, *266*, 368–375.

39. Bailey, S.E.; Olin, T.J.; Bricka, R.; Adrian, D. A review of potentially low-cost sorbents for heavy metals. *Water Res.* **1999**, *33*, 2469–2479. [CrossRef]

40. Babel, S.; Kurniawan, T.A. Low-cost adsorbents for heavy metals up take from contaminated water: A review. *J. Hazard. Mater.* **2003**, *97*, 219–243. [CrossRef]

41. Kobya, M.; Demirbas, E.; Senturk, E.; Ince, M. Adsorption of heavy metal ions from aqueous solutions by activated carbon prepared from a pricot stone. *Bioresour. Technol.* **2005**, *96*, 1518–1521. [CrossRef] [PubMed]

42. Ban, A.; Schäfer, A.; Wendt, H. Fundamentals of electrosorption on activated carbon for wastewater treatment of industrial effluents. *J. Appl. Electrochem.* **1998**, *28*, 227–236. [CrossRef]

43. López-Bernabeu, S.; Ruiz-Rosas, R.; Quijada, C.; Montilla, F.; Morallón, E. Enhanced removal of 8-quinolinecarboxylic acid in an activated carbon cloth by electroadsorption in aqueous solution. *Chemosphere* **2016**, *144*, 982–988. [CrossRef]

44. Ghosh, A.; Lee, Y.H. Carbon-Based Electrochemical Capacitors. *ChemSusChem* **2012**, *5*, 480–499. [CrossRef]

45. Noked, M.; Avraham, E.; Soffer, A.; Aurbach, D. The rate determining step of electroadsorption processes into na-noporous carbon electrodes related to water desalination. *J. Phys. Chem. C* **2009**, *113*, 21319–21327. [CrossRef]

46. Eisinger, R.S.; Alkire, R.C. Electrosorption of β-naphthol on graphite. *J. Electroanal. Chem.* **1980**, *112*, 327–337. [CrossRef]

47. Alkire, R.C.; Eisinger, R.S. Separation by Electrosorption of Organic Compounds in a Flow-Through Porous Electrode: I. Mathematical Model for One-Dimensional Geometry. *J. Electrochem. Soc.* **1983**, *130*, 85–93. [CrossRef]

48. Gabelich, C.J.; Tran, T.D.; Suffet, I.H. "Mel" Electrosorption of Inorganic Salts from Aqueous Solution Using Carbon Aerogels. *Environ. Sci. Technol.* **2002**, *36*, 3010–3019. [CrossRef]

49. Afkhami, A. Adsorption and electrosorption of nitrate and nitrite on high-area carbon cloth: An approach to purification of water and waste-water samples. *Carbon* **2003**, *41*, 1309–1328. [CrossRef]

50. Cazorla-Amorós, D.; Alcañiz-Monge, J.; De La Casa-Lillo, M.A.; Linares-Solano, A. CO2As an Adsorptive to Characterize Carbon Molecular Sieves and Activated Carbons. *Langmuir* **1998**, *14*, 4589–4596. [CrossRef]

51. Berenguer, R.; Marco-Lozar, J.; Quijada, C.; Cazorla-Amorós, D.; Morallón, E. Electrochemical regeneration and porosity recovery of phenol-saturated granular activated carbon in an alkaline medium. *Carbon* **2010**, *48*, 2734–2745. [CrossRef]

52. Beralus, J.-M.; Ruiz-Rosas, R.; Cazorla-Amorós, D.; Morallón, E. Electroadsorption of Arsenic from Natural Water in Granular Activated Carbon. *Front. Mater.* **2014**, *1*, 28. [CrossRef]

53. Berenguer, R.; Lozar, J.P.M.; Quijada, C.; Cazorla-Amorós, D.; Morallon, E. Effect of electrochemical treatments on the surface chemistry of activated carbon. *Carbon* **2009**, *47*, 1018–1027. [CrossRef]

54. Moon, H.; KookLee, W. Intraparticle diffusion inliquid phase adsorp tion of phenols with activated carbon infinite batch adsorber. *J. Colloid Interface Sci.* **1983**, *96*, 162–171. [CrossRef]

55. Yang, X.; Al-Duri, B. Kinetic model in gof liquid phase adsorption bofre active dyeson activated carbon. *J. Colloid Interface Sci.* **2005**, *287*, 25–34. [CrossRef] [PubMed]

Article

Expansion and Headloss Dependencies for Flowrate and Fluidization during Backwashing of Sand, Anthracite and Filtralite® Expanded Aluminosilicate Filters

Jaran Raymond Wood *, Tone Storbråten and Timo Neubauer

Leca International, 2009 Nordby, Norway; tone.storbraten@leca.no (T.S.); timo.neubauer@leca.no (T.N.)
* Correspondence: jaranraymond.wood@saint-gobain.com

Received: 29 August 2020; Accepted: 2 October 2020; Published: 8 October 2020

Abstract: The backwash expansion rates and headloss evolution of single- and dual-media granular filters of Filtralite® expanded aluminosilicate clay were compared with fine and coarser sand, as well as anthracite. Filtralite is manufactured in Norway, Årnesvegen 1, N-2009 Nordby. Abbreviations used for Filtralite is; N = Normal density, H = High density, C = Crushed. Each material had different particle densities and grain size distributions. The scope of the investigation was narrow: a clean-bed test was executed once for each parameter on single samples. As temperature affects the viscosity of water, tests were carried out within two temperature ranges (13–17 °C and 21–26 °C), and the effect on the fluidization of the materials was observed. The trial established that although the three types of materials have different physical properties, the expansion behaviors generally correlate with the grain sizes and particle densities of the media. To reach the expansion target of 15%, sand 1.2–2.0 mm (particle density 2656 kg/m^3) required a flow rate of 67 m/h, Filtralite HC 0.8–1.6 (1742 kg/m^3) required 34 m/h, and anthracite 0.8–1.6 mm (1355 kg/m^3) required 15 m/h. The headloss peaks that indicate fluidization were found to correspond with the onset of expansion with increasing flow rate. This was for the example observed by fluidization of 0.4–0.6 mm sand (particle density 2698 kg/m^3) at 0.94 m/m, fluidization of Filtralite HC 0.5–1 (1873 kg/m^3) at 0.46 m/m and anthracite 0.8–1.6 mm (1355 kg/m^3) at 0.21 m/m. Tests of dual-media filters of two types of Filtralite, i.e., Mono Multi and Mono Multi Fine, were also included. The backwash column used for the experiment consisted of extruded acrylic pipes with digital pressure sensors, an electronic flowmeter, a stepless pump and a water cycling system. A laminar water flow was provided by a mesh and a diffusor fixed above a single nozzle. No air was used. The trial was comparative, and its purpose was to shed light on the required water flow rates needed to fully expand different materials, and hence indicate requirements for performing proper filter backwashes.

Keywords: expanded clay; Filtralite; Mono Multi; sand; anthracite; granular filter media; backwash; headloss; clean-bed expansion; bed fluidization; water treatment

---

1. Introduction

Granular filtration is a traditional water treatment step, and sand and anthracite are classical media, but alternative products are also used. For example, Filtralite, an expanded aluminosilicate clay, has been in use worldwide for more than 30 years [1].

Filtralite is made by the high temperature treatment (~1200 °C) of certain raw clays, and this results in strong and porous grains. This process allows control over physical properties, such as size and density. The filters can be further tailored by the crushing and sieving of the expanded grains into defined fractions. The crushing of the grains opens up the macroporous structure of the expanded clay, affects the sphericity-factor, and exposes a specific surface area (SSA) of up to 23 m^2/g [2].

The large specific surface of Filtralite available for biofilm growth makes it an ideal carrier in biological treatment processes [3]. Testing of Filtralite NC 0.8–1.6 (through standard EN 1097–6, Annex E) shows a porosity of 62%, which is higher than sand or anthracite and provides extra storage capacity for suspended solids, which results in slower pressure build-ups during operations [4]. This trait makes Filtralite usable for direct filtration and coagulated water filtration processes [5].

When comparing filter run-times to reach terminal headloss, it has been found that expanded clays of 1.2 mm size can operate 2.3 times longer than anthracite, while expanded clays of 1.0 mm size or GAC (Granular Activated Carbon) can achieve 1.5 times longer filter run times than anthracite. When considering headloss development, the use of expanded clays such as Filtralite can extend overall filter runs two- to four-fold when compared with anthracite [6].

In relation to the durability of the materials, Phillip D. Davies analyzed results from different attrition tests, demonstrating that Filtralite material presents a volume loss 4.25 times lower than sand, 3 times lower than limestone and 5.5 times lower than slate [7].

By using expanded clays, increased production stability and longer production cycles could translate into potential savings in terms of lower energy and water requirements with respect to filter operations [8]. For each treatment plant, the backwashing processes has to be understood independently before applying a new filter. The data from clean-bed backwashing with novel materials such as Filtralite should be available as performance indicators prior to full-scale testing. The objectives of this trial were to show how grain size and particle densities can affect the expansion and headloss regimes for clean-bed granular filter materials of sand, anthracite and expanded clay, as well as to observe the effect on viscosity of temperature variations.

This study was initiated to provide research for operators that assesses the requirements for proper backwashes to achieve correct filter cleaning, avoid filter media losses and optimize energy use.

2. Materials and Methods

2.1. Materials

A selection of Filtralite, sand and anthracite with various densities and grain sizes was made. The characteristics are given in Table 1.

Table 1. The filter materials, their size ranges and density.

Filter Product	Fraction [1] (mm)	Particle Density [2] (kg/m³)	Bulk Density [2], (kg/m³) Compressed
Filtralite Pure HC 0.5–1	0.4–1.0	1874	779
Filtralite Pure HC 0.8–1.6	0.8–1.6	1742	712
Filtralite Pure NC 0.8–1.6	0.8–1.6	1250	422
Filtralite Pure NC 1.5–2.5	1.5–2.5	1042	424
Filtralite Pure Mono Multi	50% NC 1.5–2.5 50% HC 0.8–1.6	n/a	n/a
Filtralite Pure Mono Multi Fine	50% NC 0.8–1.6 50% HC 0.5–1	n/a	n/a
Filter-sand 1	0.4–0.6	2698	1554
Filter-sand 2	1.2–2.0	2656	1508
Anthracite	0.8–1.6 [3]	1355	698

[1] As stated by manufacturer [1], [2] single sample measured in lab, [3] 7% undersized material.

For Filtralite, the particle densities range from 1050 kg/m³ to 1800 kg/m³, and Filtralite can consequently be used in both single- and dual-media set-ups [9]. For sand and anthracite, the densities vary, but are respectively around 2600 kg/m³ and 1400 kg/m³. The typical effective sizes are for sand 0.3–0.7 mm and are for anthracite 0.7–1.7 mm [10]. This was the basis for the selection of sand grain size, which were accordingly Rådasand 0.4–0.6 mm and anthracite 0.8–1.6 mm. In order to compare sand to Filtralite 1.5–2.5, the Rådasand 1.2–2 mm was chosen.

Mono Multi and Mono Multi Fine are composed of 50–50 mixtures of normal and high-density Filtralite that constitutes a dual-media filter that can be used as an alternative to anthracite–sand dual-filters [11]. Alternatively, replacing anthracite, Filtralite can act in combination with sand in a dual-media filter set-up [12].

Filtralite appears as brown–grey, crushed grains with a sharp geometry and an open, porous internal structure. The anthracite was silvery grey, with a similar grain size distribution to that of the Filtralite 0.8–1.6. "Rådasand" is quartz filter-sand. Figure 1a–d show photos of selected materials.

(a)   (b)

(c)   (d)

Figure 1. (**a**) Filtralite Pure HC 0.5–1, (**b**) Filtralite Pure NC 1.5–2.5, (**c**) anthracite 0.8–1.6 mm, (**d**) sand 1.2–2.0 mm "Rådasand".

2.2. Methods

British Water test standard BW:P.18.93 First revision was the main guideline when carrying out the bed expansion experiments [13]. Apparent particle densities were measured by standard EN 1097-6, Annex E. Effective size D_{10} was measured by standard EN 12905. As described and illustrated in Figure 2, a transparent filter-column was constructed.

A bottom nozzle was covered with a mesh floor and a diffuser to prevent jetting. Outlets for pressure sensors (manometers) were connected to the tubing. The bottom pressure sensor outlet was placed just above the mesh, i.e., at the bottom of the filter media. The top measuring manometer was above the material bed and remained above the fluidized material during the measurements. The internal column diameter was 150 mm. Prior to the tests, all materials were submerged in water for 2 weeks to be completely saturated. There were no floating grains.

Tap-water at two different temperatures was used for the tests; nominally room-temperature water and cold, fresh tap water. Temperatures were registered by a thermocouple in the inlet water chamber. Since the setup lacked precise temperature control, data were grouped in the following ranges: (1) 13–17 °C and (2) 21–26 °C, see Tables 2 and 3. The filters were too loosely packed after initial gravitational settling, so all samples were completely compacted by knocking on the column until the material could not be further compacted. This was done to ensure consistent initial states, and the procedure is explained in a test method for filter materials [14]. Filtralite compacts very readily using this method. The depth of all filter materials in the compacted state was about 500 mm. Figure 3 below shows photos of the filter column.

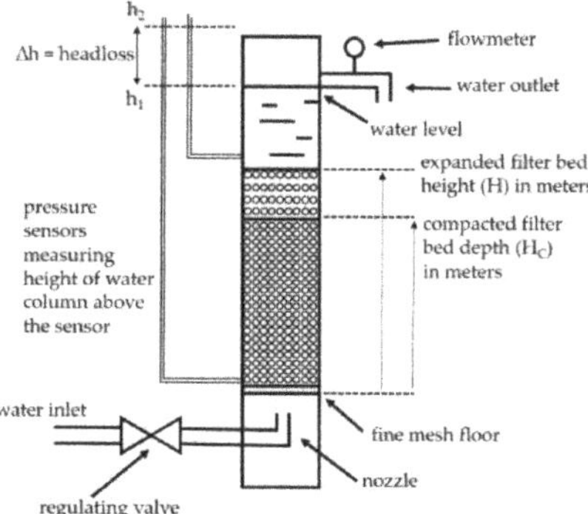

Figure 2. Principal illustration of the filter column and the placement of sensors for measuring the water pressures for the headloss calculations.

Table 2. Sizes and densities compared with constant 15% expansion flow rate.

Filter-Product	Particle Density (kg/m^3)	Effective Size, D$_{10}$	15% Expansion Flow Rate (m/h), 13–17 °C	15% Expansion Flow Rate (m/h), 21–26 °C
HC 0.5–1	1874	0.5 [1]	14	20
HC 0.8–1.6	1742	0.9 [1]	34	41
NC 0.8–1.6	1250	0.8 [1]	22	25
NC 1.5–2.5	1042	1.4 [1]	38	42
Mono Multi	n/a	n/a	38	43
Mono Multi Fine	n/a	n/a	18	27
Filter-sand 0.4–0.6	2698	0.3	22	24
Filter-sand 1.2–2.0	2656	1.2	67	73
Anthracite 0.8–1.6	1355	0.8 [2]	15	17

[1] As stated by manufacturer, [2] 7% undersized grains for anthracite.

Table 3. Size ranges and densities compared with filter headloss peaks.

Filter-Product	Particle Density (kg/m^3)	Effective Size, D$_{10}$	Headloss (m/m) at Fluidization Peak, 13–17 °C	Headloss (m/m) at Fluidization Peak, 21–26 °C
HC 0.5–1	1874	0.5 [1]	0.46	0.44
HC 0.8–1.6	1742	0.9 [1]	0.49	0.46
NC 0.8–1.6	1250	0.8 [1]	0.24	0.26
NC 1.5–2.5	1042	1.4 [1]	0.25	0.30
Mono Multi	n/a	n/a	0.40	0.39
Mono Multi Fine	n/a	n/a	0.42	0.48
Filter-sand 0.4–0.6	2698	0.3	0.94	1.00
Filter-sand 1.2–2.0	2656	1.2	0.98	0.95
Anthracite 0.8–1.6	1355	0.8 [2]	0.21	0.21

[1] As stated by the manufacturer, [2] 7% undersized grains for anthracite.

No sediments or coagulants were added, as the purpose was to investigate clean-bed expansion at different temperatures. Air was not used. All tests were firstly carried out with increasing water flow, starting at 0 m/h and ramping up with 5–15 m/h steps (dependent on media type) until the flow rate needed to obtain an expanded bed depth of 1.5 times the initial height (H_C) was reached. After full fluidization, the process was repeated, but with decreasing the rate until the surface of the materials had ceased to move. Manometer levels (h_1 and h_2) and bed height (H) were recorded for each flow rate.

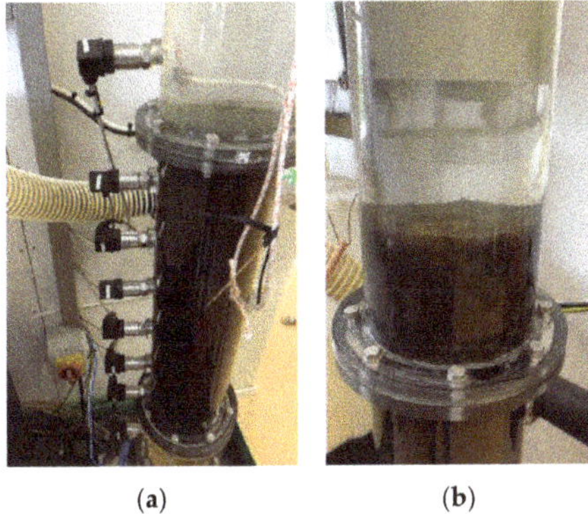

(a)   (b)

Figure 3. Photos of the filter column: (**a**) The sensor positions. During the tests, only the second sensor from the bottom and the top sensor were used; (**b**) Filtralite at about 20% fluidization during backwash.

The headloss gradient value required was the difference between the levels of the digital manometers divided by the initial (compacted) depth of the filter bed (H_C). In a basic form, headloss can be expressed as meter headloss per meter bed depth (m/m). The headloss increased with the flow rate up to the minimum fluidization flow rate (V_{mf}), where a maximum headloss (fluidization headloss plate) was reached. Headloss gradient (m/m), m headloss per m bed depth:

$$\frac{(h_2 - h_1)}{H} = m/m \tag{1}$$

The expansion (%) of the filter bed is calculated as the percentage increase in depth over the compacted depth (H_C):

$$\frac{(H - H_c)}{H_c} \times 100\% = \%_{expansion} \tag{2}$$

During backwash, friction forces act on the grains and expand the media. The state of fluidization is reached when the friction drag or the pressure drop across the filter bed is just enough to support the weight of the media [15].

Figure 4 shows a graphical illustration of the generalized expected results for headloss and expansion as a function of water flow rate (m/h). The minimum fluidization rate and fluidization headloss plate should be later recognized, as shown in this theoretical model [16].

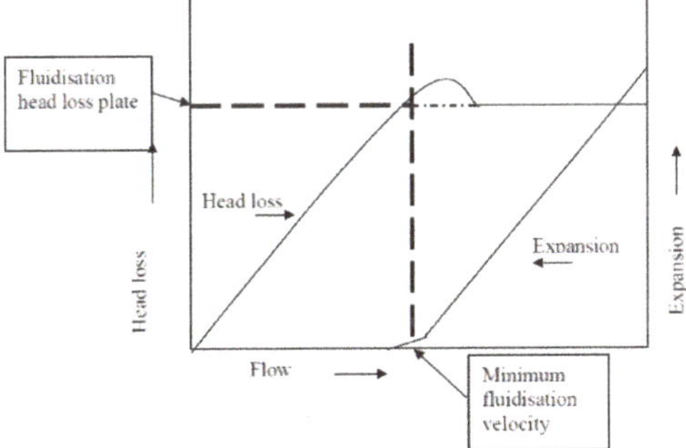

Figure 4. Illustration showing the expected theoretical expansion and headloss behavior for increasing water flow. The evolution of headloss is dependent on the point of minimum water fluidization rate and expansion.

3. Results

The results are presented in Figures 5–13. The plots represent the volume expansion (%) of the relative height difference. At 15–20% expansion, all media are in a fluidized state and scouring should be in effect [17]. The expansion target was set to be 15%, and the corresponding flow rates were recorded along with peak headlosses at partial fluidization. A 15% expansion is a typical target, as this rate usually signifies the transition from partial to full fluidization. Sphericity effects were not considered.

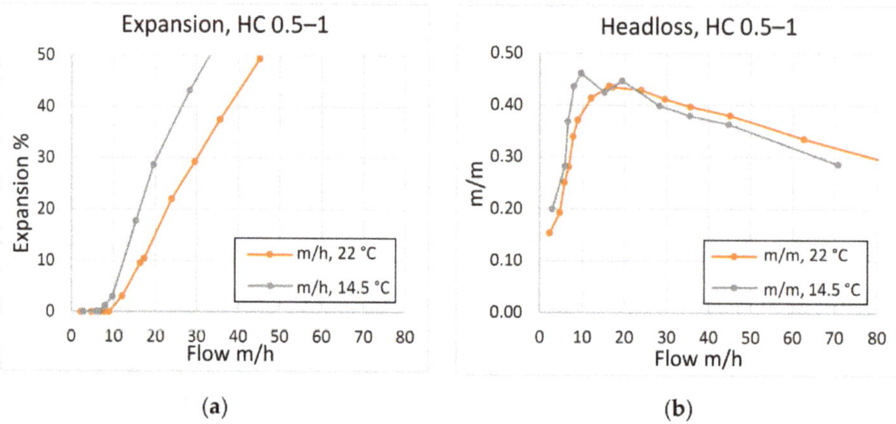

(a)　　　　　　　　　　　　　　　　(b)

Figure 5. (**a**) Expansion vs. water flow rate for Filtralite HC 0.5–1; (**b**) headloss vs. water flow rate.

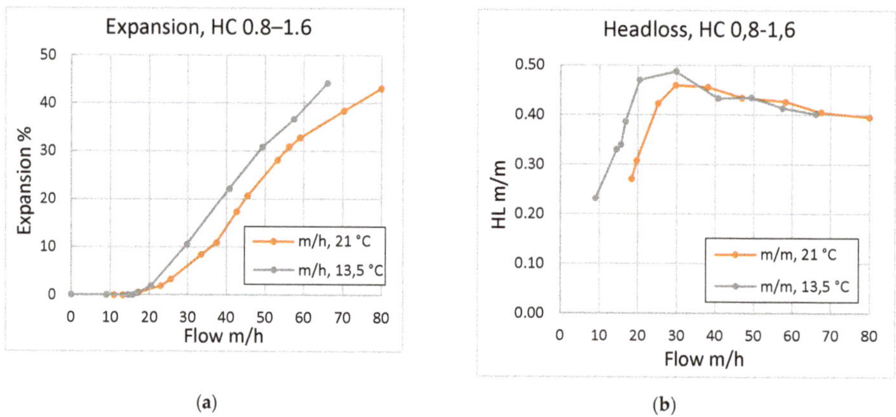

(a)　　　　　　　　　　　　　　　　(b)

Figure 6. (**a**) Expansion vs. water flow rate for Filtralite HC 0.8–1.6; (**b**) headloss vs. water flow rate.

Due to time limitations, only one test per batch of material was executed, and consequently deviations could not be recorded, although tests were performed prior to the trial to establish consistent repeatability. As recommended by standards [13,14], the measurements were done with increasing flow and decreasing flow rate. For simplicity, decreasing flow plots were omitted, as they did not significantly differ from the increasing flows plots.

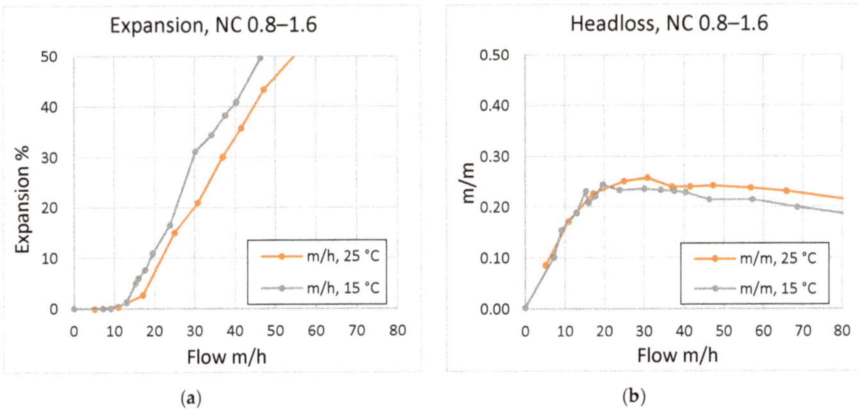

Figure 7. (**a**) Expansion vs. water flow rate for Filtralite NC 0.8–1.6; (**b**) headloss vs. water flow rate.

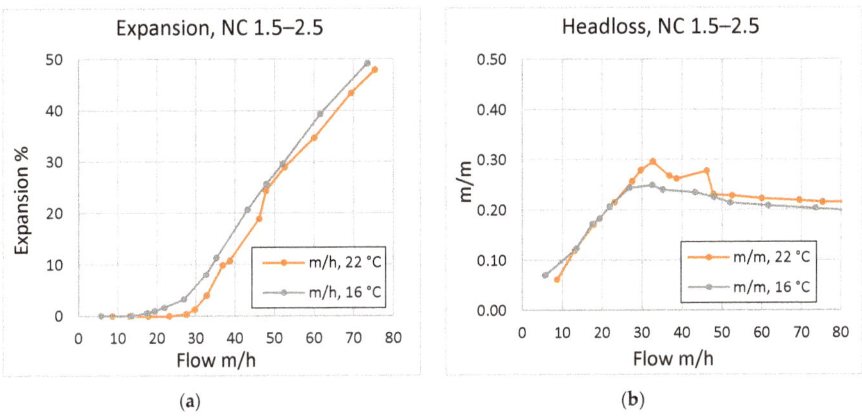

Figure 8. (**a**) Expansion vs. water flow rate for Filtralite NC 1.5–2.5; (**b**) headloss vs. water flow rate.

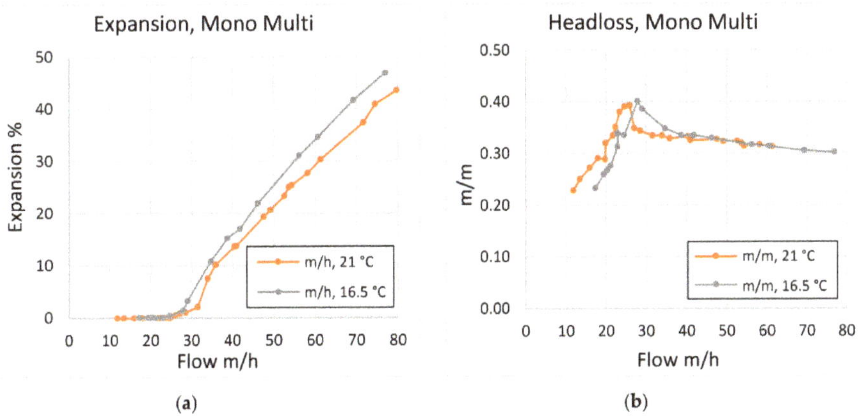

Figure 9. (**a**) Expansion vs. water flow rate for Filtralite Mono Multi; 50% NC 1.5–2.5 and 50% HC 0.8–1.6; (**b**) headloss vs. water flow rate.

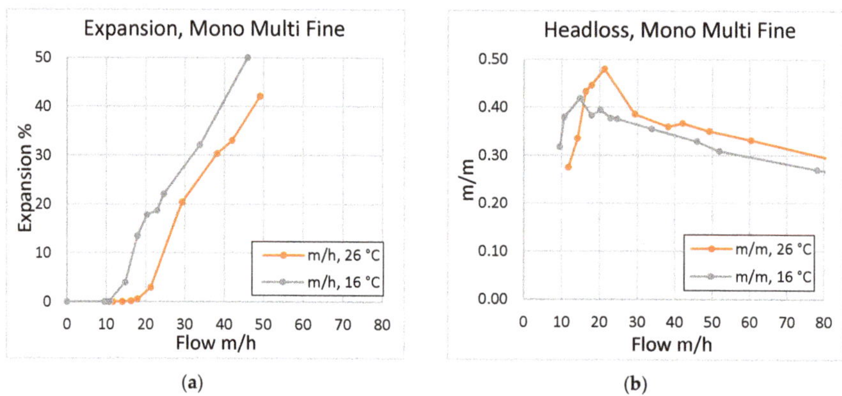

Figure 10. (**a**) Expansion vs. water flow rate for Filtralite Mono Multi Fine; 50% NC 0.8–1.6 and 50% HC 0.5–1; (**b**) headloss vs. water flow rate.

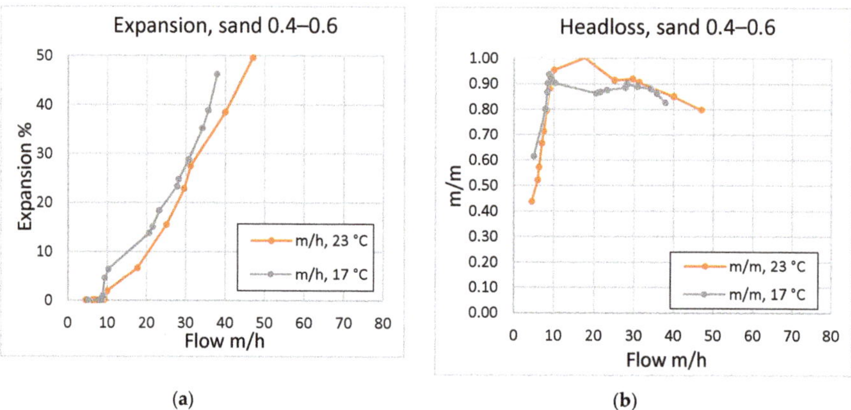

Figure 11. (**a**) Expansion vs. water flow rate for 0.4–0.6 mm filter-sand (Rådasand); (**b**) headloss vs. water flow rate.

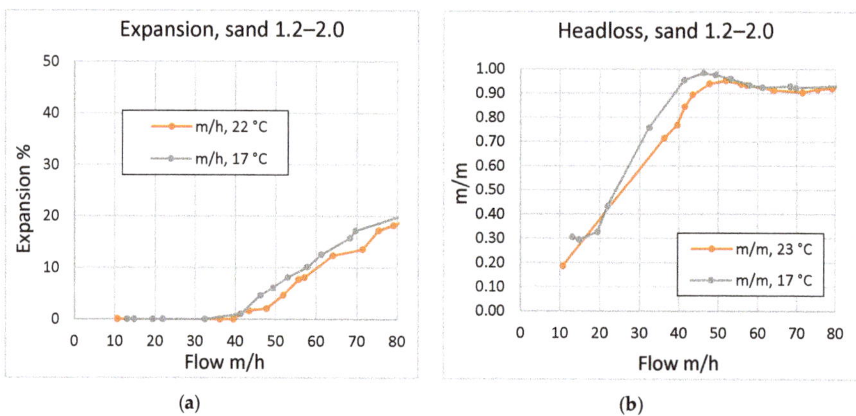

Figure 12. (**a**) Expansion vs. water flow rate for 1.2–2.0 mm filter-sand (Rådasand); (**b**) headloss vs. water flow rate.

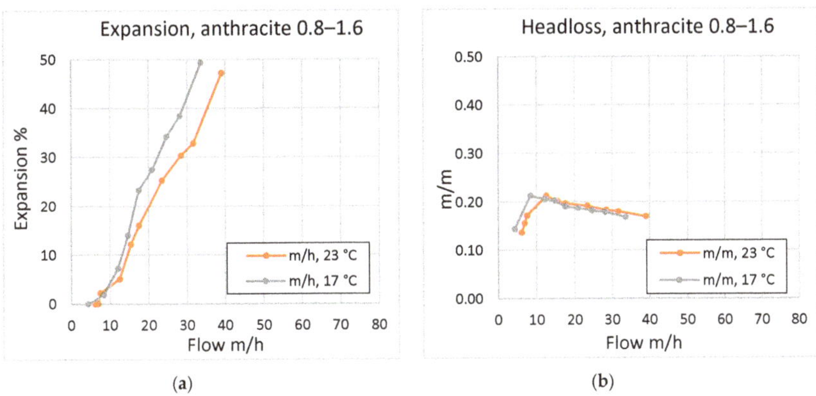

Figure 13. (**a**) Expansion vs. water flow rate for anthracite 0.8–1.6 mm; (**b**) headloss vs. water flow rate.

4. Discussion

Selected backwashing properties of common granular filter media were analyzed according to standardized tests [13,14].

Firstly, the purpose of the experiments was to compare filter materials with different physical properties, and secondly, to indicate the dependence of backwash expansion on water temperature. Raw water may seasonally vary above and below 20 °C, hence temperature conditions are highly relevant to treatment plants that need to adapt water production to viscosity.

The overall results suggest that both filter density and media size can be adjusted to suit backwash regimes and operational limitations for a unique treatment train. Increasing grain size and porosity will act towards maintaining hydraulic permeability, and the filters will resist high pressure drops during backwashes, especially when compacted. Furthermore, the finer grained materials expand more readily as a result of the induced pressure, i.e., headloss, acting on the grains during backwash.

Listed in Table 2 are the corresponding flow rates necessary for 15% expansion in the two temperature ranges. Partial fluidization is sufficient to cause the rubbing of grains, whilst avoiding full fluidization, as full fluidization provides negligible rubbing and may cause filter wash-out [17].

A key consideration is media particle grain size distribution and size range. Comparisons of effective sizes and fluidization headlosses at different temperatures are summarized in Table 3.

In the dissertation of T. Saltnes (2002), a linear increase in headloss with an increasing flow rate before fluidization, as well as a constant headloss pattern after bed expansion, are seen [4]. Saltnes observed that the fluidization of anthracite–sand starts at about 12 m/h, and at about 28 m/h for the Filtralite filter, a result that correlates fairly well with the current study.

For further reference, Table 4 provides a list of backwashing requirements from an extensive case study on granular filtration media, with regards to biological aerated filters (BAF) [18].

Table 4. Summary of biologically active filter (BAF) backwashing requirements.

	Backwash Rate (m/h)	Air Scour Rate (m/h)	Total Duration (min)
Upflow, sunken media			
Normal BW	20	97	50
Energetic BW	30	97	25
Downflow, sunken media	15	90	20–50

The final backwashing requirements and duration are often developed via cooperation between designer and user. For example, the backwash sequence for an upflow sunken media BAF can include drain down, air scour, air and water scour (and may also include cycling between air only and air/water scour), water-only rinse, and a filter-to-waste step when the filter bed is placed back in operation, and the whole process is adjusted during long-term operation [18]. It follows that backwash rates are highly dependent on the configuration of a treatment train and the filter-bed hardware, and must be considered when assessing plant-specific data on the performance of a new material or a specific filtering process for a full-scale pilot test.

Possible continuations of this study could be abrasion, settling rates, turbidity performance and filter run-times for ripening and breakthrough, as well theoretical calculations for examining further dependencies.

5. Conclusions

It was established that differences in fluidization rates could be attributed to density and grain sizes, as also seen in earlier research [7].

For the two established temperature ranges (13–17 °C and 21–26 °C), the temperature dependency of water viscosity could be observed, as all expansion rates were higher at the lower temperatures.

A clear temperature dependency of headloss at the fluidization plate was not demonstrated, as the fluidization headlosses were not consistently suppressed at higher temperatures, and vice a versa.

To illustrate the dependencies, the following conclusions refer to the temperature range 13–17 °C.

The corresponding flow rates for achieving the expansion target (15%) increased with particle density. This was seen when comparing sand 1.2–2.0 mm (2656 kg/m^3) with a 67 m/h flow rate, HC 0.8–1.6 (1742 kg/m^3) with a 34 m/h flow rate, and anthracite 0.8–1.6 mm (1355 kg/m^3) with a 15 m/h flow rate.

Anthracite 0.8–1.6 had lower headloss and expansion rate (respectively, 15 m/h and 0.21 m/m vs. 22 m/h and 0.24 m/m for the slightly lower particle density of 1250 kg/m^3 for NC 0.8–1.6). This conflicts with the hypothesis, and the effect may be due to the high amount of undersized grains (7%) in the anthracite, over-packing in the column, or an unaccounted-for characteristic of anthracite.

Sand 1.2–2 (2656 kg/m^3), with its higher particle density compared with the less dense NC 1.5–2.5 (1042 kg/m^3), required a higher flow rate to achieve the expansion target—respectively 67 m/h and 38 m/h. This supports the conclusion that increasing particle density requires increasing the corresponding flow rate to reach the expansion target.

The fine sand 0.4–0.6 had an early expansion at 22 m/h, but at the cost of high headloss (0.94 m/m) due to its narrow grain size distribution. As HC 0.5–1 (1874 kg/m^3) and sand 0.4–0.6 (2698 kg/m^3) both have comparatively early expansion regimes, respectively 14 m/h and 22 m/h. Between the materials, the main difference was the headloss plate, as the sands 0.4–0.6 vs. HC 0.5–1 differed in performance by 0.46 m/m and 0.94 m/m, respectively. Thus, it can be concluded that lower fluidization headlosses can be attributed to particle density as well [19].

The expansion regimes of the Filtralite dual-layer combinations Mono Multi and Mono Multi Fine performed as expected when intermediately compared with their single-layer components—respectively 38 m/h and 18 m/h—and presumably operated with the added benefit of increased solids capacity [20].

Author Contributions: Conceptualization, J.R.W. and T.S.; methodology, T.S.; software, J.R.W.; validation, J.R.W.; formal analysis, J.R.W. and T.N.; investigation, J.R.W. and T.N.; resources, J.R.W.; data curation, J.R.W.; writing—original draft preparation, J.R.W.; writing—review and editing, J.R.W.; visualization, J.R.W.; supervision, J.R.W.; project administration, J.R.W. All authors have read and agree to the published version of the manuscript.

Funding: This research received no external funding.

Conflicts of Interest: The authors declare no conflict of interest.

References

1. Filtralite Manufacturer's Web-Site. Available online: www.filtralite.com (accessed on 31 August 2020).
2. Blom, R.; Spjelkavik, A.I.; Kvello, J. *SINTEF Materials & Chemistry; BET* Analyses of Various Products and Pre-Products; *(Brunauer–Emmett–Teller); Project no. 102019024; 02.10.2018;* SINTEF Materials and Chemistry: Oslo, Norway, 2018.

3. García-Ruiz, M.; Maza-Márquez, P.; González-Martínez, A.; Campos, E.; González-López, J.; Osorio, F. Performance and bacterial community structure in three autotrophic submerged biofilters operated under different conditions. *J. Chem. Technol. Biotechnol.* **2018**, *93*, 2429–2439. [CrossRef]

4. Saltnes, T. Contact Filtration of Humic Waters in Expanded Clay Aggregate Filters. Ph.D. Thesis, Norwegian University of Science and Technology, Trondheim, Norway, March 2002.

5. Ødegaard, H.; Østerhus, S.; Melin, E.; Eikebrokk, B. NOM removal technologies-Norwegian experiences. *Drink. Water Eng. Sci.* **2010**, *3*, 1–9. [CrossRef]

6. Sharma, D. Taylor-Edmonds L, Andrews R C; Comparative assessment of ceramic media for drinking water biofiltration. *Water Res.* **2018**, *128*, 1–9. [CrossRef] [PubMed]

7. Davies, D.P. Alternative Filter Media in Rapid Gravity Filtration of Potable Water. Ph.D. Thesis, Institutional Repository, Loughborough University, Loughborough, UK, 2012.

8. Uhl, W.; Nahrstedt, A. Research Report; Investigation of the Newly Developed Filter Media Filtralite on Its General Filtration Performance in Surface Water Treatment for Drinking Water and with Special Respect to Removal of Microbial Contaminants; University of Duiburg-Essen; Germany; Institute for Energy- and Environmental Process Engineering Water Technology and IWW Rheinisch-Westfälisches Institut für Wasser Mülheim, Ruhr. 30 June 2004. Available online: https://filtralite.com/sites/default/files/pdfs/Investigation_of_filtralite_general_filtration_performance_-_iww_report.pdf (accessed on 17 August 2020).

9. Saltnes, T.; Eikebrokk, B.; Ødegaard, H. Contact filtration of humic waters: Performance of an expanded clay aggregate filter (Filtralite) compared to a dual anthracite/sand filter. *Water* **2002**, *2*, 17–23. [CrossRef]

10. SUEZ Handbook of Industrial Water Treatment; Chapter 06 Filtration. Available online: www.suezwatertechnologies.com/handbook/chapter-06-filtration (accessed on 31 August 2020).

11. Steele, M.E.; Chipps, M.J.; Bayley, R.; Mikol, A.; Fitzpatrick, C.S.B. Alternative filter media for potable water treatment; Article; Thames Water Research & Development; Department of Civil & Environmental Engineering University College London, 05.01.2016; Research Report. Available online: https://issuu.com/e-weber/docs/alternative_filter_media_for_potabl (accessed on 31 August 2020).

12. Eikebrokk, B.; Haaland, S.; Jarvis, P.; Riise, G.; Vogt, R.D.; Zahlsen, K. *NOMiNOR: Natural Organic Matter in Drinking Waters within the Nordic Region*; Norwegian Water Report 231/2018; Norwegian Water BA: Vangsvegen, Norway, 2018; ISBN 978-82-414-0406-1.

13. British Water Test Standard. BW:P.18.93 First revision. In *Standard for the Specification, Approval and Testing of Granular Filtering Materials*; British Water; 1 Queen Anne's Gate: London, UK, 1996; ISBN 0-9509979-6X.

14. *NF X 45-402:1996-12; Matériax filtrants; E: Granular Filtering Media—Test Methods—Functional Characteristics*; AFNOR: Paris, France, 1996; Available online: https://www.boutique.afnor.org/standard/nf-x45-402/granular-filtering-media-test-methods-functional-characteristics/article/780428/fa043059 (accessed on 17 August 2020).

15. Cleasby, J.L.; Rice, G.A.; Stangl, E.W.; McKeown, G.H. Development in backwashing of granular filters. *J. Environ. Eng. Div.* **1975**, *101*, 713–727.

16. Fan, K.S. Sphericity and Fluidization of Granular Filter Media. Master's Thesis, Iowa State University, Ames, IA, USA, 1981.

17. Kawamura, S. Design and Operation of High-Rate Filters-Part 2. *J. Am. Water Works Assoc.* **1975**, *67*, 653–662. [CrossRef]

18. Debarbadillo, C.; Rogalla, F.; Tarallo, S.; Boltz, J.P. Factors Affecting the Design and Operation of Biologically Active Filters. In Proceedings of the Water Environment Federation, Portland, OR, USA, 15–18 August 2010. [CrossRef]

19. Mitrouli, S.T.; Karabelas, A.J.; Yiantsiosa, S.G.; Kjølseth, P.A. New granular materials for dual-media filtration of seawater: Pilot testing. *Sep. Purif. Technol.* **2009**, *65*, 147–155. [CrossRef]

20. Dobiáš, P.; Dolejš, P. Filtralite FMMF pilot and full-scale operation experience. In *Conference Papers "Pitná Voda 2016"*; W&ET Team, Budejovice, C., Eds.; Poloprovozní a provozní zkušenosti s použitím; Filtralite–FMMF: Tábor, Czech Republic, 2016; pp. 81–86. ISBN 978-80-905238-2-12014.

© 2020 by the authors. Licensee MDPI, Basel, Switzerland. This article is an open access article distributed under the terms and conditions of the Creative Commons Attribution (CC BY) license (http://creativecommons.org/licenses/by/4.0/).

Article

WWTP Effluent Quality Improvement for Agricultural Reuse Using an Autonomous Prototype

Laura Ponce-Robles [1,*], Beatriz Masdemont-Hernández [2], Teresa Munuera-Pérez [2],
Aránzazu Pagán-Muñoz [3], Andrés Jesús Lara-Guillén [3], Antonio José García-García [1],
Francisco Pedrero-Salcedo [1], Pedro Antonio Nortes-Tortosa [1] and Juan José Alarcón-Cabañero [1]

[1]   Department of Irrigation, Centro de Edafología y Biología Aplicada del Segura, CEBAS-CSIC,
      30100 Murcia, Spain; antonio.j.garcia2@gmail.com (A.J.G.-G.); fpedrero@cebas.csic.es (F.P.-S.);
      panortes@cebas.csic.es (P.A.N.-T.); jalarcon@cebas.csic.es (J.J.A.-C.)
[2]   SISTEMA AZUD, S.A., 30820 Murcia, Spain; beatriz@azud.com (B.M.-H.); tmunuera@azud.com (T.M.-P.)
[3]   Technology Centre for Energy and the Environment (CETENMA), 30353 Murcia, Spain;
      aranzazu.pagan@cetenma.es (A.P.-M.); alara@cetenma.es (A.J.L.-G.)
*   Correspondence: lponce@cebas.csic.es; Tel.: +34-968-396-200

Received: 5 July 2020; Accepted: 5 August 2020; Published: 9 August 2020

Abstract: Wastewater reuse presents a promising way to mitigate the risk to global water resources and achieve sustainability in water, especially in agricultural areas in the southeast of Spain, such as the Murcia region. However, the risks related to the presence of contaminants of emerging concern (CECs) or pathogenic microorganisms in wastewater treatment plant (WWTP) effluent suggest the need to implement effective and relatively low-cost tertiary treatments. With this aim, a self-sustainable pilot prototype based on three combined modules (disc-filtration, granular activated carbon (GAC) bed adsorption and UV disinfection) assisted by solar panels was installed as an alternative tertiary treatment in a conventional WWTP in the Murcia region. The obtained results clearly confirmed the efficiency of the proposed prototype for CECs removal, and showed optimal results at a workflow of 500 L/h. In all cases, high removal efficiency was obtained for the different indicator microorganisms described in the recently published Regulation (EU) 2020/741 (*E. coli*, F-specific coliphages, somatic coliphages, total coliphages, and *Clostridium perfringens*). The protection of the activated carbon by disc-filters and the energy autonomy and self-operation of the prototype resulted in an efficient and economically viable methodology for its implementation in both conventional WWTPs and in isolated areas attached to crops.

Keywords: granular activated carbon; adsorption; autonomous prototype; emerging contaminants; filtration; reclaimed water; safe agricultural reuse

---

1. Introduction

Nowadays, the use of reclaimed water is seen as an excellent approach to deal with water scarcity. Indeed, water reuse is an important feature of water policies in most countries around the world, for example, it is included in WHO guidelines [1], and it is seen as a fundamental aspect of the concept of a circular economy [2]. In the European Union, one billion cubic meters of urban wastewater are reused annually (approximately 2.4% of treated effluents). However, this value is expected to increase by up to six times due to the development of numerous initiatives regarding water reuse for irrigation, industrial uses or aquifer recharge [3]. This practice not only promotes the efficient use of water, but also leads to prosperity from an environmental, human and economic point of view, and promotes management strategies for the safe discharge of wastewater into the environment and protection of surface and groundwater [4]. Possible reuse applications include agricultural or landscape irrigation, groundwater recharge, industrial activities, street cleaning or ecological uses [5,6].

In particular, in semi-arid areas of the Mediterranean basin, where agriculture is the highest consumer of water (approximately 68% in Spain, 50% in Italy and 80% in Greece), the use of reclaimed water for agricultural purposes is essential [7–9]. Water reuse also has important economic and environmental benefits, such as the possibility of nutrient recovery or reduction in fertilizer use [10,11]. However, the use of reclaimed water in agriculture requires quality control to guarantee satisfactory sanitary conditions. Therefore, one of the current goals is the development and improvement of technologies that enable the adequate treatment of wastewater [12] and to achieve the discharge limits established for agricultural uses according to the current and future legislation. Specifically, at the European level, the recently published Regulation (EU) 2020/741 of the European Parliament and of the Council of 25 May 2020 (https://eur-lex.europa.eu/legal-content/EN/TXT/?uri=uriserv%3AOJ.L_.2020.177.01.0032.01.ENG&toc=OJ%3AL%3A2020%3A177%3ATOC) on minimum requirements for water reuse establishes the minimum quality requirements to use reclaimed water for agricultural purposes, to ensure a high level of protection of the environment and human and animal health, and to promote the circular economy [13]. The Commission estimates that new regulations could increase water reuse in agricultural irrigation from 1.7 billion m^3 to 6.6 billion m^3 per year, thereby reducing water stress by 5% (EU, 2018 estimations) [14]. However, special attention has recently been given to so-called "contaminants of emerging concern" (CECs), which are mostly unregulated compounds that may be candidates for future regulation, depending on research on their potential effects on health and data that monitors their occurrence. The list of CECs includes a wide variety of products for daily use in both industrial and domestic applications, such as pharmaceuticals and personal care products (PPCPs), hormones, surfactants, endocrine disruptors, antiseptics, pesticides and synthetic fragrances [15,16].

One of the main sources of water for potential agricultural reuse comes from municipal wastewater treatment plants (WWTPs). However, conventional secondary (usually activated sludge processes) and tertiary (media filtration and disinfection) wastewater treatments are not effective in the removal of CECs [17–19]. Several authors have studied the removal efficiencies of different CECs in conventional WWTP using activated sludge. Behera et al. reported low removal values for carbamazepine (23.1%), atenolol (64.5%), metoprolol (23%), sulfamethazine (13.1%) and sulfamethoxazole (51.9 %) [20], while Radjenovic et al. reported low removal values for compounds such as diclofenac (50.1%), indomethacin (23.4%), mefenamic acid (29.4%), ofloxacin (23.8%) and erythromycin (23.8%) [21]. The presence of CECs in WWTPs effluent due to low removal efficiencies is of particular relevance in agricultural reuse because of the possibility of CECs uptake and accumulation in food crops and consequent diffusion into the food chain. Although some CECs are already among the priority research lines of the main organizations dedicated to the protection of public and environmental health, such as WHO and environmental protection agencies such as the Environmental Protection Agency (EPA), no legal discharge limits are described in the current legislation. However, due to the growing interest in these substances, some of them, such as diclofenac and erythromycin are included in the European Union Watch List (Directive, 2013/39/EU, modified by Decision 2015/495/EU of 20 March 2015 and updated in Decision 2018/840/EU of 5 June 2018) and discharge limits will probably be included in future regulations [22–24].

With that in mind, the implementation of tertiary treatments in real WWTPs specifically designed for CECs removal will be crucial to efficiently obtain safe, high-quality reclaimed water. This will increase the trust of farmers and the final consumers of the products and simplify the proper management of the water. However, the problems that these compounds may generate in the short or long term, the lack of specific legislation obliging WWTPs to use such treatments, along with other economic problems (investment costs, maintenance, energy consumption, the need for high-complexity analytics) make it difficult to implement the appropriate treatments in WWTPs. Other factors that may limit the implementation of specific technologies in WWTPs include the great diversity of existing CECs, and the different physico-chemical properties that they have.

Recent studies have focused on several treatment processes applied for the removal of CECs such as advanced biological processes (membrane bioreactors and bio-electrochemical systems) [25,26], and advanced oxidation processes (AOPs) [27,28]. The main limitations of these types of technologies are their high operating costs, specific infrastructures and high energy consumption. Other alternatives such as activated carbon adsorption and ozonation are considered as economically feasible options that can be coupled to WWPTs [29], however, the use of ozone may be associated with the generation of by-products, which can be hazardous to health if the water is used for agricultural purposes.

In this context, the adsorption of CECs by granular activated carbon (GAC) is a promising technology due to its removal efficiency at relatively low cost [30]. In general, and due to its properties (such as its high porosity, high surface area and high degree of surface interactions), activated carbon has a high adsorption capacity for a wide spectrum of CECs [31,32]. In addition, this process has the advantage of not generating secondary by-products [32,33]. Although the effectiveness of adsorption processes in CECs removal is well-known, the introduction of tertiary treatments in conventional WWTPs for the purpose of safe water for agricultural reuse is poorly implemented. This is mainly due to a lack of knowledge about adsorption mechanisms, the great diversity of CECs present in wastewater, the large amount of wastewater interference, and lack of knowledge about the design of systems capable of removing CECs that guarantee the production of CEC-free effluent at relatively low costs.

The aim of this work was the integration of a self-sustainable pilot plant prototype in a conventional WWTP for the purpose of safe agricultural reuse. The design focused on the removal of CECs and microbiological risks by using a combined Filtration-GAC-UV system, in line with current and future legislation. The use of water from a real WWTP together with data on the efficiency and associated economic costs will give us a practical view of the implementation of such systems.

2. Materials and Methods

2.1. Wastewater Samples

The water source used in the experiments for this study was a WWTP located in the Murcia region (Spain), (latitude 37°47′48″ N, longitude 0°57′36″ W). This plant receives domestic and industrial effluent from three municipalities (Roldán, Lo Ferro and Balsicas) and has a maximum treatment capacity of 2,007,500 m^3/year of wastewater.

This WWTP consists of a secondary treatment based on active sludge with extended aeration and secondary sedimentation. It also uses a tertiary treatment based on coagulation-flocculation, rapid sand filter and disinfection using ultraviolet (UV) radiation. Treated wastewater is commonly used for agricultural irrigation in the Campo de Cartagena area.

2.2. Chemicals and Reagents

Seven CECs that have been detected worldwide in WWTP influents and effluents (Tran et al., 2018) were selected for the study: (acetaminophen (ACT), carbamazepine (CBZ), diclofenac (DCF), erythromycin (ERY), ketoprofen (KTP), naproxen (NPX) and sulfamethoxazole (SMX)) [16]. Sigma-Aldrich (Steinheim, Germany) analytical standards of these compounds at high purity grade (>99%) were used. The physico-chemical properties and chemical structure of these CECs are given in Table 1.

Table 1. Physico-chemical properties and chemical structure of selected contaminants of emerging concern (CECs).

Compound	Structure	Mw (g/mol)	Water Solubility (mg/L)	Log Kow	pKa
Acetaminophen (ACT) NSAIDs [1]	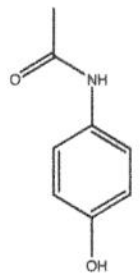	151.17	14,000 (at 25 °C)	0.46	9.38
Carbamazepine (CBZ) Anticonvulsant	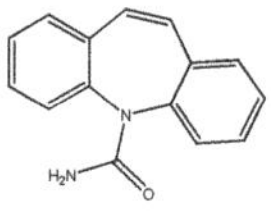	236.27	18 (at 25 °C)	2.45	13.9
Diclofenac (DCF) NSAIDs [1]	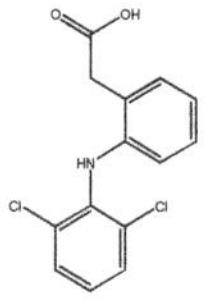	296.15	2.37 (at 25 °C)	4.51	4.15
Erytromycin (ERY) Antibiotic	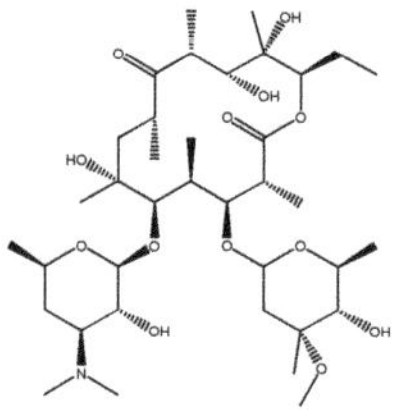	733.94	4.2 (at 25 °C)	3.06	8.88
Ketoprofen (KTP) NSAIDs [1]	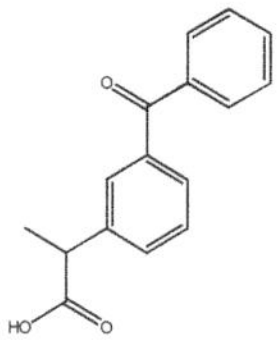	254.29	51 (at 22 °C)	3.12	4.45
Naproxen (NPX) NSAIDs [1]	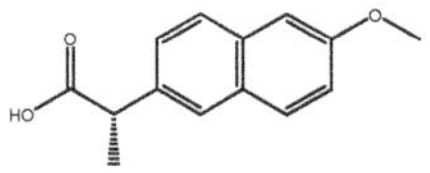	230.26	15.9 (at 25 °C)	3.18	4.15
Sulfamethoxazole (SMX) Antibiotic	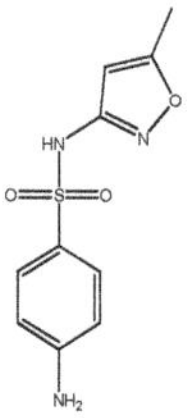	253.27	610 (at 37 °C)	0.89	1.6/5.7

[1] NSAIDs: Nonsteroidal anti-inflammatory drugs.

A commercial coconut shell-based GAC with high density value (520 ± 30 kg/m^3) was used as adsorbent in accordance with Benstoem et al. [34]. In that paper, the authors described a high correlation between the high density of GAC sorbents and their capacity to adsorb CECs.

The selected GAC was activated thermally with steam and supplied by Chiemivall (Castellbisbal, Barcelona, Spain). The main properties are described in Table S1, Supplementary Materials.

2.3. Analytical Determinations

Wastewater samples were characterized in terms of pH, turbidity, electrical conductivity (EC), dissolved organic carbon (DOC), total suspended solids (TSS), five-day biochemical oxygen demand (BOD5), and some ions (for conditions, see Appendix A).

The microbiological analysis of the CECs (ACT, CBZ, DCF, ERY, KTP, NPX and SMX) (*Clostridium perfringens* spores, *Escherichia Coli* and total coliphages (somatic and F-specific)) was done by an external ISO 17025 certified laboratory (IPROMA S.L.) located in Castellón, Spain. For CECs, SPE extraction followed by HPLC-MS (High-performance liquid chromatography – mass spectrometry) was used. *Clostridium perfringens* spores and *Escherichia Coli* were determined by plate count while coliphages (total, somatic and F-specific) were determined using the double agar layer method (ENAC n° 109/le285 accreditation, https://www.enac.es/documents/7020/ac514131-7717-42cc-b026-0b8de53c78e1.

2.4. Preliminary Lab-Scale Experiments

The adsorption capacity of the selected GAC was studied at lab scale using a methacrylate column of 550 mm length and 19 mm internal diameter containing 21.5 g of GAC. The experiments were carried out using secondary WWTP effluents fortified with 200 µg/L of each CEC. The mixture was added to a plastic tank and pumped into the column by a peristaltic pump (P-Selecta, Percom-I model, (J.P. Selecta, Abrera, Barcelona, Spain) Operational parameters (Table 2) were selected according to previous results described in the literature [35], which were also compatible with the dimensions established for the design of the experimental prototype.

Table 2. Operational parameters of the adsorption lab-scale preliminary experiments.

Parameter	Units
GAC bed height	20 cm
Empty bed contact time	39.38 s
Filtration area of GAC	2.8 cm^2
Flow rate	5.18 L/h
Filtration velocity	18.28 m/h
Volume of water needed	145 L

Effluent was collected from the lab-system at different time intervals until the concentration remained constant over time. The amount of adsorbed CEC was calculated as follows [36,37]:

$$q_e = \frac{(C_0 - C_i) \, x \, V}{m} \tag{1}$$

where C_0 is the initial CEC concentration (µg/L), C_i is the CEC concentration in treated wastewater (µg/L), V is the volume of treated wastewater (L) and m is the mass of GAC (g). All adsorption experiments were performed at least twice and the results are reported as an average.

2.5. Plant Prototype Design

The prototype (Figure 1) was designed, manufactured and installed by AZUD in the CEBAS-CSIC's Research Platform inside the Roldán-Lo Ferro-Balsicas WWTP, and it was directly connected to effluent from the secondary treatment.

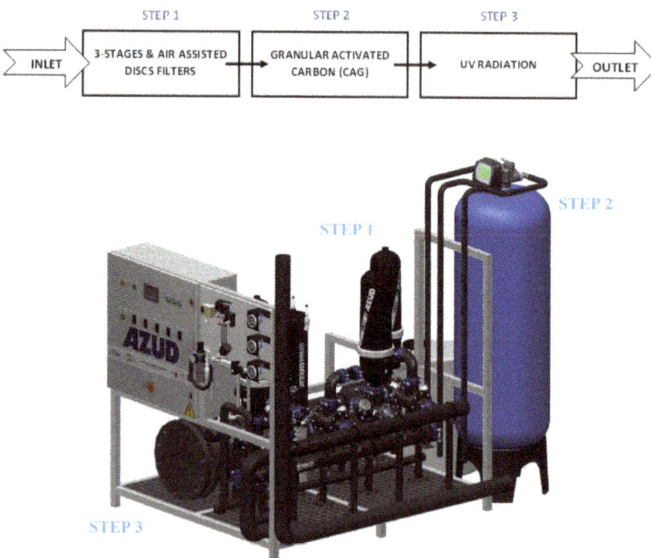

Figure 1. Prototype design.

The maximum workflow of the prototype is 1200 L/h although hydraulically it is capable of working with lower flows (minimum 500 L/h). The water treatment includes three steps:

1   The first step is based on disc filters AZUD HELIX AUTOMATIC®.

The objective of the first step is to filter the water out of the secondary settling tanks in order to allow the next step of the treatment (GAC) to work specifically on the adsorption of the CECs without any interference due to suspended particles that might reduce the adsorption performance.

Disc filters were selected to achieve this because this allows for a deep filtration that eliminates all particles larger than the degree of filtration and a high percentage of smaller ones, regardless of the geometry or nature of the particles. This is also an effective way of protecting the rear media filter. The disc filter module has a specific design that combines three independent stages working in series with several selected particle retention capacities (50, 20 and 5 μm). This experimental module includes an automatic backwash system that combines water and compressed air, which allow for the disc recovery. The air-assisted backwash system (AA) works with low pressure (even during the backwash process), thus reducing the hydraulic interaction in the prototype during cleaning and optimizing the water and energy consumption as well as obtaining total recovery of the discs. The backwashing times was 30 s for each stage.

2   The second step is based on the adsorption process using a high-quality coconut shell-based GAC.

A total of 112 kg of CG900 (Chiemivall, Castellbisbal, Barcelona, Spain), (see Table S1, Supplementary Materials) was placed in a bed column made of polyester reinforced with fiberglass of 1945 mm high with an auto-control valve. Weekly cleaning of the module was scheduled to ensure maximum repetitiveness of the analytical results.

3   The third step is based on a system for disinfection with UV radiation. This step was designed for disinfection purposes and consists of a conventional λ = 245 nm UV lamp with 200–400 W that was self-assembled in the prototype.

The prototype is totally autonomous due to the automation of all processes. Five solar panels (270POL, Sunconnection WorldWide 2016®, Murcia, Spain) were installed outside to provide the required electric energy (Figure 2a). The panels supply a total of 1650 W, while the prototype consumes energy in a range between 840 W and 1600 W depending on the selected configuration (For more details, see Table S2). Solar panels were connected to the electric feed of the system through a set of batteries (Figure 2b) for the continuous operation of the system even when the solar radiation is low or during the hours without sun.

Figure 2. (**a**) Solar panels (the lined ones in the lower part of the picture); (**b**) Set of batteries.

3. Results and Discussion

3.1. Characterization of WWTP Effluent

Samples from the secondary effluent from the WWTP were characterized between March and June 2019 in order to assess the variability in real effluents. Table 3 shows the average, minimum and maximum values of the different physico-chemical parameters analyzed. In general, low standard deviations were obtained for almost all parameters analyzed, indicating effluent stability. The greatest variations in standard deviation were found in the analysis of Cl^- and SO_4^{2-}, which coincided with samples taken from point of entry to the WWTP.

Table 3. Characteristics of secondary wastewater treatment plant effluents ($n = 20$).

Parameters	Minimum	Maximum	Average	SD
pH	7.19	7.73	7.43	0.15
Turbidity (NTU)	0.95	7.79	2.66	1.60
Electrical Conductivity (EC, dS/m)	1.11	1.80	1.57	0.15
5-day Biochemical Oxygen Demand (BOD_5, mg/L)	3.00	12.00	6.25	4.03
Total Nitrogen (mg/L)	6.55	11.74	9.42	1.82
Dissolved organic carbon (mg/L)	9.76	13.90	11.67	1.39
Total Solids (TSS, mg/L)	3.00	10.00	5.50	2.00
Transmittance (%)	48.80	64.70	53.69	4.58
F^- (mg/L)	<0.02	0.14	0.18	0.03
Cl^- (mg/L)	141.72	301.02	243.57	47.13
NO_2^- (mg/L)	<0.10	1.44	1.38	0.08
Br^- (mg/L)	0.26	0.51	0.38	0.08
NO_3^- (mg/L)	2.45	19.74	9.93	5.74
PO_4^{3-} (mg/L)	5.96	14.67	9.08	2.91
SO_4^{2-} (mg/L)	136.00	231.03	200.82	30.91
SAR	2.83	5.59	4.90	1.01

3.2. Adsorption Capacity of Selected GAC

Before the commissioning of the prototype, the GAC maximum adsorption capacity was determined at lab-scale. The results (reported as the average concentration of all selected CECs) showed that after 120 L of wastewater was treated, the concentration remains constant, so, with regard to Equation (1), the maximum adsorption capacity was determined as 0.8 mg CECs/g GAC. Therefore, a maximum adsorption of 89.6 g of CECs can be expected in the GAC bed column of the self-sustainable prototype.

3.3. Prototype Evaluation and Optimization

3.3.1. Disc-Filter Module

To evaluate the operation and efficiency of the disc-filter module, the pressure differential was continuously monitored, since this parameter allows evaluation of the degree of filter fouling. A conservative maximum pressure value for each of the selected filters (50, 20 and 5 μm) was set at 0.6 bars. When this value was reached, the automatic backwashing process was activated. The results (Figure 3) showed that of the total volume of prototype influent, only 0.23% was used for cleaning the filters, while 8.9% was destined to a second module cleaning (GAC bed column), and the water obtained was 90.87% of the total volume.

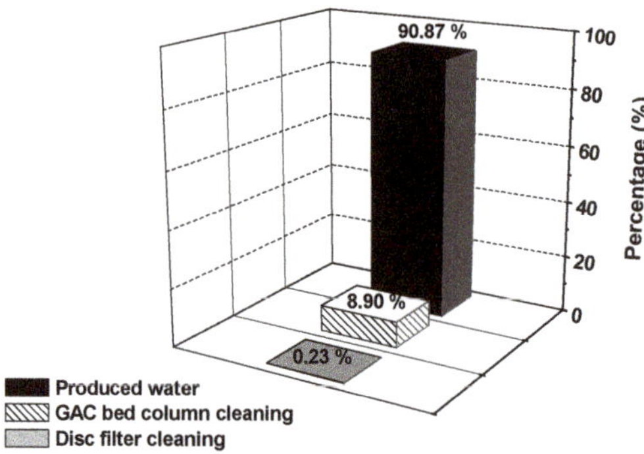

Figure 3. Percentages of produced water, granular activated carbon (GAC) bed column cleaning and disc filter cleaning from total treated volume.

Our results are in accord with Ravina et al. [38], who suggested that by using disc filters, the volume of the backwash water was generally less than 0.5% of the total volume passing though the filter. However, considering that this low percentage corresponds to the cleaning of the filter with the smallest particle size (5 μm), the number of backwashes with respect to the accumulated treated volume was studied in order to evaluate the long-term efficiencies of the module (Figure 4). A total volume of 1000 m^3 of real secondary WWTP effluents was passed through the filtering module. The variability in the frequency of cleaning is mainly due to the TSS that can enter the filters from secondary WWTP effluents according to Duran-Ros et al. [39], however, in almost all cases, values under 20 were obtained. The trend followed a virtually horizontal line, indicating that the incidence of cleaning remained constant over time, which suggests that the system is able to recover its initial state. In general, system fouling is related to the cleaning frequency, so that if the cleaning frequency increases, the filtering capacity decreases, until the filter clogs. For a detailed investigation of disc filter fouling, it would be necessary to do a study with much larger volumes or to work with effluents containing higher TSS values (as shown in Table 3, the selected effluents have low TSS values).

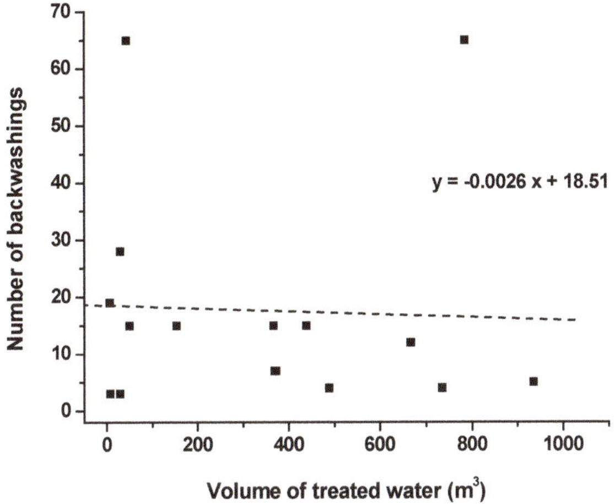

Figure 4. Number of automatic backwashes performed with respect to the treated volume.

3.3.2. GAC Bed Column Module

- Flow rate effects on CECs removal efficiency

In general, the retention time of CECs in the bed column is one of the most important factors affecting the efficiency of adsorption, due to its influence on both the adsorbate and adsorbent. The variation in empty bed contact time (EBCT) in the GAC bed column was studied through the variation in the workflow rate. The maximum and minimum prototype workflow (1200 and 500 L/h), and three intermediate points (1000, 800, and 600 L/h) were studied. The values of the EBCT that corresponded with these points are detailed in Table 4. All experiments were performed by introducing real secondary WWTP effluents into a 5000 L mixing tank where additional quantities of each CECs (200 µg/L) were added. In order to obtain representative results, experiments were performed during several months (September and November 2019 and February 2020) and all analytical samples were analyzed in duplicate. The overall results are shown in Table 4.

Table 4. CECs, dissolved organic carbon (DOC), turbidity and transmittance values obtained at different workflows ($n = 6$).

Workflow (L/h)	EBCT (min)	HLR * (m/h)	Total CECs Removal (%)	DOC Removal (%)	Turbidity Removal (%)	Transmittance Increase (%)
1200	11	5.3	52.5 ± 2.2	22.3 ± 2.1	16.1 ± 2.0	15.5 ± 3.9
1000	14	4.4	53.4 ± 4.3	23.8 ± 3.0	18.0 ± 3.1	16.0 ± 4.6
800	17	3.5	54.6 ± 1.6	24.3 ± 2.8	19.0 ± 2.9	18.0 ± 3.8
600	23	2.6	62.3 ± 4.9	25.6 ± 3.1	21.0 ± 4.1	21.5 ± 4.2
500	27	2.2	74.6 ± 6.1	27.2 ± 4.0	27.5 ± 3.8	23.5 ± 5.6

* HLR: Hydraulic loading rate.

In all experiments, pH values remained constant at neutral pH (7.1–7.9). The selected CECs were moderately to effectively removed over the entire course of the experiments, with mean (±standard deviation) removals ranging from 52.5% ± 2.2% to 74.6% ± 6.1%. Results showed that CECs removal increased with the EBCT values, indicating that lower flow rates are most effective for CECs adsorption compared with higher flow rates. The same tendency occurred with DOC and turbidity removals, while an increase of % transmittance was detected. According to Rahman et al., this is due to two phenomena: (i) lower workflow rates imply higher residence time in the bed column, and (ii) compounds or ions present in wastewaters have more time to diffuse into the pores of GAC through intra-particle diffusion [40]. Therefore, in order to obtain high-quality effluents for agricultural uses, a workflow of 500 L/h (EBCT 27 min) showed the best adsorption efficiencies. Under these conditions, the ability to remove individual CECs was evaluated.

The results (Figure 5) showed that the adsorption capacity for individual CECs followed the following trend: ACT > CBZ > NPX > KTP > DCF > ERY > SMX. The best adsorption results (95.77 ± 4.3%) were obtained for ACT mainly due to its low molecular weight compared to the rest of the CECs and its aromatic structure (see Table 1). In general, adsorption capacity is strongly related to the physical-chemical properties of CECs. Those with smaller molecular weights and aromatic compounds are more susceptible to adsorption in GAC systems, while higher molecular weight or low aromatic compounds are poorly adsorbed by GAC [41,42].

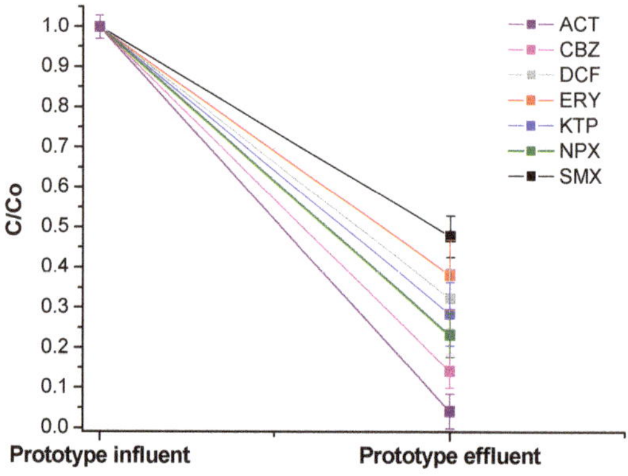

Figure 5. Adsorption efficiencies of selected CECs in workflow of 500 L/h.

Specifically, ERY has a significantly higher molecular weight than the rest (733.94 g/mol). Furthermore, it is not aromatic, which has proven to be a disadvantageous adsorption characteristic, and it showed lower adsorption values in comparison with the other compounds (61.81% ± 8.61%).

The behavior of SMX is unusual compared to the above—it showed the lowest adsorption capacity (52.24% ± 5.15%). This may be related to its reactivity in aqueous solution, which results in two dissociation constants [43]. Because of this, SMX may be present in solutions in anionic form. Anionic species have a low tendency to adsorption by GAC, and may also be displaced by other adsorbates according to Yang et al. [44]. The low adsorption efficiency of SMX in GAC systems has been reported by several authors. For example, Motwani et al. reported that only 27.27% of SMX was removed using GAC with a particle size of 2.0–5.0 mm, while none or little adsorption were described by Telgmann et al. using different GAC sorbents (8 × 30 mesh, 4 × 8 mesh and 8 × 14 mesh of particle size) [45,46].

- Inlet CECs concentration effects

Similar to previous studies, the effects of inlet CECs concentration were studied using real secondary WWTP effluents. Additional quantities of each CEC (at two different concentrations that differ by almost an order of magnitude, 200 and 50 µg/L) were introduced in a 5000 L mixing tank, and spiked effluents were passed through the GAC bed column at a workflow of 500 L/h. All analytical samples were analyzed in triplicate and the results are reported in Figure 6.

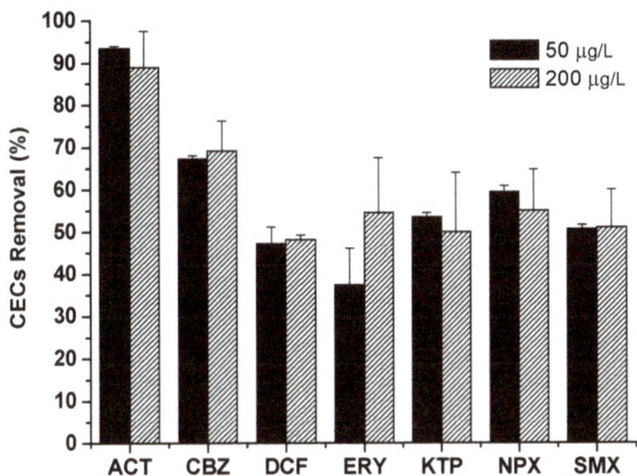

Figure 6. CECs removal percentages at two inlet concentrations (200 and 50 μg/L).

With regard to the high variability of real WWTP secondary effluents, no significant effects were found between two studied CECs concentrations, except for ERY, which increased from 37.29 ± 8.76% to 54.55 ± 12.86% with concentrations of 50 and 200 μg/L, respectively. This can be explained by considering that at high initial concentrations, the ratio of ERY molecules to adsorbent GAC active sites is high, therefore, the probability of the interaction of molecules increases. Similar effects of inlet concentration are common in GAC systems and these have been included in several studies. Sotelo et al. [47] reported an increase in the percentage of caffeine (from 77.71% to 91.74%) and diclofenac (from 58.75% to 71.54%) adsorption when the initial concentration of these compounds increased from 3 to 7 mg/L, while Ang et al. [48] reported an increase in sevoflurane (an anesthetic) adsorption (from 536 ± 5 mg/g to 604 ± 9 mg/g and from 329 ± 5 to 368 ± 5 mg/g) when two different GACs (E-GAC and H-GAC, respectively) were used at inlet concentrations from 55.9 and 527.9 mg/L.

3.3.3. UV-Module Efficiency

Removal of pathogenic microorganisms is essential to guarantee the high quality of prototype effluents. Therefore, the efficiency of the selected UV-lamp was evaluated by monitoring different indicator microorganisms as described in the recently published Regulation (EU) 2020/741 of the European Parliament and of the Council of 25 May 2020 on minimum requirements for water reuse (E. coli for pathogenic bacteria, F-specific coliphages, somatic coliphages and total coliphages for pathogenic viruses, and Clostridium perfringens spores for protozoa) [13]. The monitoring of these indicators in influent and effluent prototype samples was carried out for a period of one year (March 2019–March 2020) and the values are shown in Table 5. The maximum and minimum prototype workflow (1200 and 500 L/h) was also evaluated.

Table 5. Monitoring of microbiological indicators in the prototype at two workflows (500 and 1200 L/h) during a period of one year (*n* = 40).

Indicator	Prototype Influent Values			% Removal	
	Minimum	Maximum	Mean	Flow 500 L/h	Flow 1200 L/h
Clostridium perfringens spores (CFU/100 mL)	310	16,000	3000	100	99.9
Escherichia coli (CFU/100 mL)	770	35,000	9484	100	99.9
Total Coliphages (MPN/100 mL)	2	220	47	100	100
Somatic Coliphages (MPN/100 mL)	<1	210	45	100	100
F-Specific Coliphages (MPN/100 mL)	<1	8	3	100	100

Clostridium perfringens spores and Escherichia coli were detected in all prototype influent samples with average values of 3000 and 9484 CFU/100 mL, respectively, while lower values were detected for Coliphages (the average values for total, somatic and F-Specific were 47, 45 and 3 MPN/100 mL, respectively).

At a workflow of 500 L/h, all microbiological parameters were removed, while an average 0.1% decrease in Clostridium perfringens spores and Escherichia coli was detected for workflows of 1200 L/h. Despite this small loss of efficiency, high-quality effluents were obtained in all cases according to (EU) 2020/741 regulations [13], and they also complied with the most restrictive crop category values (Class A: All food crops consumed raw where the edible part is in direct contact with reclaimed water and root crops consumed raw).

Therefore, and independently of the workflow, high-quality effluents for reuse purposes were obtained, which minimized the microbiological risk of ingesting any type of crop irrigated with reclaimed waters.

3.3.4. Continuous Operation of the Prototype

Considering the previous results for CECs and microbiological indicators removal, the workflow was adjusted to 500 L/h and the continuous operational work mode was evaluated. The work time was set for 8 h of work per day and weekly samples were evaluated for several months (March 2019–2020). At each sampling, both the influents (samples from the secondary treatment of the WWTP) and the prototype effluents were analyzed.

Analysis of the CECs in the prototype influent showed low concentration values for the selected compounds, with a minimum and maximum of 0.4 µg/L (DCF) and 3.20 µg/L (KTP), while ACT and ERY were not found in any analyzed sample with values below the analytical limits (limit of quantitation (LOQ)) marked by the reference laboratory that analyzed the samples (Table 6). The physico-chemical and microbiological analysis of prototype influents showed values within the ranges in all cases as previously described in Tables 1 and 5.

Table 6. CECs concentration in the prototype influent.

CECs	LOQ (µg/L)	Prototype Influent Concentration (µg/L)	
		Minimum	Maximum
ACT	0.4	<LOQ	<LOQ
CBZ	0.2	<LOQ	0.53
DCF	0.2	0.40	1.30
ERY	0.2	<LOQ	<LOQ
KTP	0.2	<LOQ	3.20
NPX	0.4	<LOQ	0.60
SMX	0.2	<LOQ	0.40

Prototype effluent samples showed average values for turbidity removal (28.58% ± 5.15%) and transmittance increase (20.66% ± 2.30%) similar to those obtained in previously reported values for the GAC module (Table 4), so, the variation in these parameters is directly related to the GAC efficiency. On the other hand, an average removal of 43.36% ± 12.14% of DOC was found in the prototype effluent. This is partly due to the efficiency of organic matter disposal in the GAC step (approximately 27.2% ± 4.0%), and the rest is due to its retention in the first stage (disc filters).

In all cases, the values obtained in prototype effluents for turbidity, TSS and BOD_5 were lower than the most restrictive values reported in the recently published (EU) 2020/741 Regulation on minimum requirements for water reuse (turbidity ≤ 5 NTU, TS ≤ 10 mg/L and BOD_5 ≤ 5 mg/L), while EC and SAR were lower than 3 dS/m and 6, respectively, which is in accord with Spanish regulations (Royal Decree 1620/2007, https://www.boe.es/buscar/doc.php?id=BOE-A-2007-21092) [49]. In addition, microbiological indicators showed the total removal of selected parameters (according to Table 5), while values below the analytical limits set (Table 6) for the selected CECs were found in all cases, indicating high-quality agronomic effluents, suitable for safe, agricultural reuse purposes.

After the good results obtained, and taking into account the low concentrations of CECs found in the prototype influent and its stability, as shown in Table 1, an additional study using WWTP secondary effluents was carried out to evaluate the binomial high quality vs. volume of produced water, with a flow rate of 1200 L/h (maximum prototype workflow). The results were similar to those obtained for a 500 L/h workflow and showed high efficiency in terms of CECs removal with lower LOQ values in all cases, while the physico-chemical and microbiological indicators were below the standard limits indicated above. These results indicate that higher workflows could be used with medium-high quality water, while in water with a high concentration of CECs such as industrial wastewaters (which may also contain high values of turbidity or organic matter) it is necessary to work at low flows to ensure the high agronomic quality of the effluent. Thus, previous characterization of WWTP effluents is necessary.

Although high quality effluents were obtained in both minimum and maximum workflows (500 and 1200 L/h) in this particular case, after treatment of 715.33 m^3 of wastewater, traces of CECs were found in prototype effluents (1200 L/h workflow). In particular, between 0.5 μg/L to 1 μg/L of CBZ, DCF, KTP, and SMX were detected, suggesting the saturation of the GAC column.

For a more detailed study, the decrease in adsorption efficiency at different treated wastewater volumes was evaluated (Figure 7). In accord with the lower CECs concentration in real WWPT effluents, a few control experiments were programmed using real WWTP effluents containing selected CECs (200 μg/L of each). Different volumes of treated wastewater (330, 440 and 715 m^3) were evaluated until the detected CECs point described above. The 500 and 1200 L/h workflows were compared (Figure 7).

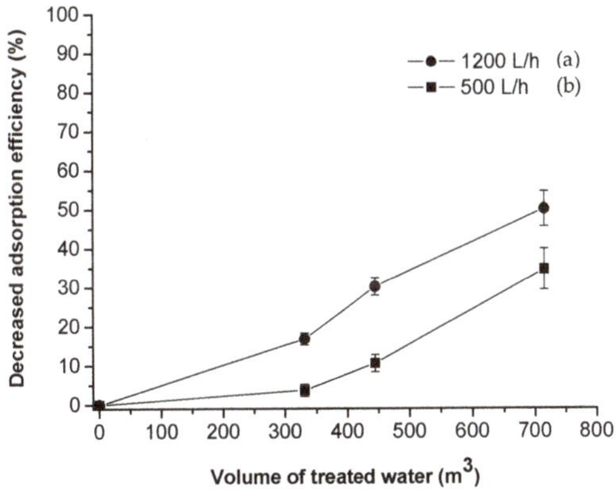

Figure 7. Decrease in adsorption efficiency at different treated wastewater volume in two workflows: (a) 1200 L/h, (b) 500 L/h.

In both cases, the adsorption capacity decreased when the treated volume increased, suggesting that adsorption capacity gradually deteriorates with use. Specifically, a 50.05% ± 4.48% and 34.58% ± 5.24% decrease in adsorption efficiency was obtained for 1200 and 500 L/h workflows at 715 m^3, respectively. The results suggest that when the adsorption capacity percentage decreases to around 50%, the GAC should be replaced.

According to Chae et al. [50], the rate at which carbon is exhausted and the frequency at which carbon should be replaced/regenerated can be calculated by determining the carbon usage rate (CUR), which is the mass of activated carbon used per unit volume of wastewater treated (Equation (2)).

$$CUR = \frac{mass\ of\ GAC\ in\ column}{treated\ volume} \tag{2}$$

Therefore, according to this equation, and selecting 715.33 m^3 as the maximum safety volume (conservative value for all possible prototype workflows), the CUR value obtained would be 0.15 kg of GAC per m^3 of treated volume (112 kg/715.33 m^3), which can be used as a predictive value for scaled-up GAC systems.

3.4. Applicability and Cost Assessment

Although several technical or environmental factors have to be taken into account in the selection or feasibility of implementing an effective tertiary system, cost is often the determining factor. Taking into account the advantages of the selected GAC in the adsorption of CECs, we considered two possible scenarios.

The first scenario involves the implementation of the prototype in isolated areas, where an improvement in water quality is necessary prior to irrigating crops, in a conventional WWTP that does not have an installed tertiary treatment. An estimation of the operation and maintenance costs of the prototype (excluding installation) are shown in Table 7. Considering 8 h of work per day and 365 days/year (2920 h/year), the theoretical cost is 1300 €/year. On the other hand, if we consider the possibility of working at the maximum flow rate of the prototype (1200 L/h), the expected treated volume of water is 3504 m^3/year. However, since the treated volume is approximately 90% of the total entry to the prototype (see Figure 3), the treated volume of water would be 3154 m^3/year. This means a maintenance cost of 0.37 €/m^3 treated or 0.41 €/m^3 produced, which is a relatively acceptable value given that the average cost of reused water treatment in Spain is about 0.46 €/m^3 for a system that includes a physico-chemical operation, membrane filtration, reverse osmosis and chlorination [51].

Table 7. Annual operation and maintenance cost of the prototype.

Operation Type	Amount/Year	€/Year
Maintenance [1]	26 h	390
Cleaning [1]	8 h	120
Operation [1]	16 h	240
GAC change	1 change	350
UV lamp	0.25 ud	200
Electrical energy	Solar panels	0
TOTAL		1300

[1] Estimated cost of personnel: 15 €/h.

The second scenario is the inclusion of GAC modules in a conventional WWTP that already contain filtration steps and disinfection tertiary treatments. Thus, the long-term operational data from a WWTP with a conventional design was compared with the same design including a GAC step.

In the particular case of the Murcia region, the largest urban waste water regeneration system for the production of quality regenerated water 2.1 (according to the criteria of the Royal Decree 1620/2007 (2007), and MAPAMA recommendations (2010)) includes several stages: (1) coagulation-flocculation suspended matter reduction, using polyaluminium chloride (PAC) and anionic polyelectrolyte (PA); (2) filtration by sand filters, micro-sieves or textile filters; and (3) disinfection with UV lamps, and a final dose of NaClO.

Assuming a duration of 20 years, and that inversion costs are evenly distributed over that time, the implementation of a GAC system in a conventional WWTP would increase costs from approximately 0.117 €/m^3 to 0.142 €/m^3, that is, an increase of 21%. This increase is mainly due to the costs of replacing the GAC and the increased energy consumption of installations (around 25%), while the remaining increase in costs is due to the initial investment (12%) and 5% is due to equipment and facilities maintenance.

In both scenarios, the higher costs are related to the current limited capacity to regenerate GAC (replacement costs). However, regeneration studies, together with the development of new adsorbents capable not only of purifying water but also recovering valuable substances, provide useful perspectives on the implementation of these processes.

4. Conclusions

This work clearly demonstrates the feasibility of implementing a self-sufficient prototype in conventional WWTP as a tertiary treatment of effluent for agricultural reuse purposes. Regardless of the workflow, the use of GAC provides a promising way to reduce CECs concentration, while the UV lamp was able to remove all microbiological parameters analyzed (*E. coli*, total coliphages (F-specific and somatic) and *Clostridium perfringens* spores).

Lower workflows (high EBCT) increased the CECs removal efficiency, particularly for low molecular weight and aromatic compounds. However, to select the workflow, it is necessary to know the initial quality of the water to be treated (concentration and nature of CECs, presence of interferences, etc.), as well as the required quality of the produced water.

The use of the previous filtration stage (disc filters) obtained a good proportion of filtered water versus water consumption, while the energy autonomy and self-operation of the prototype resulted in an economically viable methodology for its implementation in conventional WWTPs, and in isolated areas linked to crops.

In view of the limited information available on the large-scale use of efficient tertiary treatments for reuse purposes, the information obtained in this study will help to provide a reference for efficient industrial designs.

Supplementary Materials: The following are available online at http://www.mdpi.com/2073-4441/12/8/2240/s1, Table S1: Activated carbon CG 900 characterization, Table S2: Solar panels characterization.

Author Contributions: Investigation, L.P.-R., B.M.-H., A.P.-M., A.J.G.-G.; writing—original draft preparation, L.P.-R.; writing—review and editing, L.P.-R., B.M.-H., A.P.-M., F.P.-S.; supervision, P.A.N.-T., J.J.A.-C., A.J.L.-G. and T.M.-P. All authors have read and agreed to the published version of the manuscript.

Funding: This research was supported by the strategic project Ris3mur REUSAGUA, funded by Consejería de Empleo, Universidades, Empresa y Medio Ambiente (Government of Murcia) under the European Regional Development Fund ERDF 2014–2020, which started in September 2016 and will run until the end of December 2020.

Acknowledgments: We would like to thank all the partners involved in the current project: Centro De Edafología y Biología Aplicada Del Segura (CEBAS), AZUD, Empresa Municipal de Aguas y Saneamiento de Murcia (EMUASA), Hidrogea, Instituto Murciano de Investigación y Desarrollo Agrario y Alimentario (IMIDA), Centro Tecnológico de la Energía y Medio Ambiente (CETENMA), Entidad de Saneamiento y Depuración de Aguas Residuales de la Región de Murcia (ESAMUR), Universidad de Murcia (UMU) and Universidad Politécnica de Cartagena (UPCT).

Conflicts of Interest: The funders had no role in the design of the study; in the collection, analyses, or interpretation of data; in the writing of the manuscript, or in the decision to publish the results.

Appendix A

pH and EC were measured with a multi-parameter equipment Eutech PC 2700 (Eutech instruments, Singapore). Turbidity was analyzed with a turbidity-meter Dinko D-110 (Dinko Instruments S.A., Barcelona, Spain). TSS values were determined according to American Standard Methods (American Public Health Association American Water Works Association, Water Pollution Control Federation, and Water Environmental Federation, 1915). Dissolved organic carbon (DOC) was measured in a Shimadzu 5050 TC-TOC-TN analyzer (Shimadzu Corporation S.L. Japan). Total Dissolved Nitrogen (TN) was measured using the same TC-TOC-TN analyzer. Ions were measured by ion chromatography using a Metrohm chromatograph (Metrohm, Switzerland). Samples for DOC, TN, spectrometry and chromatography methods were previously filtered using 45 µm filters.

Sodium Adsorption Ratio (SAR) was calculated based on the relation between soluble sodium and soluble calcium and magnesium divalent cations [52]:

$$SAR = \frac{Na^+}{\sqrt{(Ca^{+2} + Mg^{+2})/2}}$$

References

1. World Health Organization. *WHO Guidelines for the Safe Use of Wasterwater Excreta and Greywater*; World Health Organization: Geneva, Switzerland, 2006; Volume 1.
2. Voulvoulis, N. Water reuse from a circular economy perspective and potential risks from an unregulated approach. *Curr. Opin. Environ. Sci. Health* **2018**, *2*, 32–45. [CrossRef]
3. Official Website of the European Union. Available online: https://ec.europa.eu/environment/water/reuse.htm (accessed on 30 July 2020).
4. Lazarova, V.; Levine, B.; Sack, J.; Cirelli, G.; Jeffrey, P.; Muntau, H.; Salgot, M.; Brissaud, F. Role of water reuse for enhancing integrated water management in Europe and Mediterranean countries. *Water Sci. Technol.* **2001**, *43*, 25–33. [CrossRef]
5. Lazarova, V. Global milestones in water reuse: Keys to success and trends in development. *Water* **2013**, *15*, 12–22.
6. Eslamian, S. *Urban Water Reuse Handbook*; CRC Press: Boca Raton, FL, USA, 2016.
7. Morari, F.; Giardini, L. Municipal wastewater treatment with vertical flow constructed wetlands for irrigation reuse. *Ecol. Eng.* **2009**, *35*, 643–653. [CrossRef]
8. Meneses, M.; Pasqualino, J.C.; Castells, F. Environmental assessment of urban wastewater reuse: Treatment alternatives and applications. *Chemosphere* **2010**, *81*, 266–272. [CrossRef]
9. Pedrero, F.; Kalavrouziotis, I.; Alarcón, J.J.; Koukoulakis, P.; Asano, T. Use of treated municipal wastewater in irrigated agriculture—Review of some practices in Spain and Greece. *Agric. Water Manag.* **2010**, *97*, 1233–1241. [CrossRef]
10. Candela, L.; Fabregat, S.; Josa, A.; Suriol, J.; Vigués, N.; Mas, J. Assessment of soil and groundwater impacts by treated urban wastewater reuse. A case study: Application in a golf course (Girona, Spain). *Sci. Total Environ.* **2007**, *374*, 26–35. [CrossRef] [PubMed]
11. Alcon, F.; Pedrero, F.; Martin-Ortega, J.; Arcas, N.; Alarcon, J.J.; De Miguel, M.D. The non-market value of reclaimed wastewater for use in agriculture: A contingent valuation approach. *Span. J. Agric. Res.* **2010**, *8*, 187–196. [CrossRef]
12. Jhansi, S.C.; Mishra, S.K. Wastewater treatment and reuse: Sustainability options. *Consilience* **2013**, *10*, 1–15.
13. Official Website of the European Union. Available online: https://eur-lex.europa.eu/eli/reg/2020/741/oj (accessed on 28 June 2020).
14. European Parliament Website. Available online: https://www.europarl.europa.eu/thinktank/en/document.html?reference=EPRS_BRI(2018)625171 (accessed on 30 July 2020).
15. Westerhoff, P.; Yoon, Y.; Snyder, S.; Wert, E. Fate of endocrine-disruptor, pharmaceutical, and personal care product chemicals during simulated drinking water treatment processes. *Environ. Sci. Technol.* **2005**, *39*, 6649–6663. [CrossRef] [PubMed]
16. Tran, N.H.; Reinhard, M.; Gin, K.Y.H. Occurrence and fate of emerging contaminants in municipal wastewater treatment plants from different geographical regions—A review. *Water Res.* **2018**, *133*, 182–207. [PubMed]
17. Li, B.; Zhang, T. Mass flows and removal of antibiotics in two municipal wastewater treatment plants. *Chemosphere* **2011**, *83*, 1284–1289. [CrossRef] [PubMed]
18. Rizzo, L.; Fiorentino, A.; Grassi, M.; Attanasio, D.; Guida, M. Advanced treatment of urban wastewater by sand filtration and graphene adsorption for wastewater reuse: Effect on a mixture of pharmaceuticals and toxicity. *J. Environ. Chem. Eng.* **2015**, *3*, 122–128. [CrossRef]
19. Krzeminski, P.; Tomei, M.C.; Karaolia, P.; Langenhoff, A.A.M.; Almeida, C.M.R.; Felis, E.; Gritten, F.; Andersen, H.R.; Fernandes, T.; Manaia, C.M.; et al. Performance of secondary wastewater treatment methods for the removal of contaminants of emerging concern implicated in crop uptake and antibiotic resistance spread: A review. *Sci. Total Environ.* **2019**, *648*, 1052–1081. [CrossRef] [PubMed]

20. Behera, S.K.; Kim, H.W.; Oh, J.E.; Park, H.S. Occurrence and removal of antibiotics, hormones and several other pharmaceuticals in wastewater treatment plants of the largest industrial city of Korea. *Sci. Total Environ.* **2011**, *409*, 4351–4360. [CrossRef]

21. Radjenovic, J.; Petrovic, M.; Barceló, D. Analysis of pharmaceuticals in wastewater and removal using a membrane bioreactor. *Anal. Bioanal.* **2007**, *387*, 1365–1377. [CrossRef]

22. Directive, E.U. Directive 2013/39/EU of the European Parliament and of the Council of 12 August 2013 amending Directives 2000/60/EC and 2008/105/EC as regards priority substances in the field of water policy. *Off. J. Eur. Union* **2013**, *226*, 1–17.

23. Barbosa, M.O.; Moreira, N.F.; Ribeiro, A.R.; Pereira, M.F.; Silva, A.M. Occurrence and removal of organic micropollutants: An overview of the watch list of EU Decision 2015/495. *Water Res.* **2016**, *94*, 257–279.

24. Jurado, A.; Walther, M.; Díaz-Cruz, S. Occurrence, fate and environmental risk assessment of the organic microcontaminants included in the Watch Lists set by EU Decisions 2015/495 and 2018/840 in the groundwater of Spain. *Sci. Total Environ.* **2019**, *663*, 285–296. [CrossRef]

25. Grandclement, C.; Seyssiecq, I.; Piram, A.; Wongwahchung, P.; Vanot, G.; Tiliacos, N.; Roche, N.; Doumenq, P. From the conventional biological wastewater treatment to hybrid processes, the evaluation of organic micropollutant removal: A review. *Water Res.* **2017**, *111*, 297–317.

26. Tiwari, B.; Sellamuthu, B.; Ouarda, Y.; Drogui, P.; Tyagi, R.D.; Buelna, G. Review on fate and mechanism of removal of pharmaceutical pollutants from wastewater using biological approach. *Bioresour. Technol.* **2017**, *224*, 1–12. [CrossRef] [PubMed]

27. Ahmed, M.B.; Zhou, J.L.; Ngo, H.H.; Guo, W.; Thomaidis, N.S.; Xu, J. Progress in the biological and chemical treatment technologies for emerging contaminant removal from wastewater: A critical review. *J. Hazard. Mater.* **2017**, *323*, 274–298. [CrossRef] [PubMed]

28. Rodriguez-Narvaez, O.M.; Peralta-Hernandez, J.M.; Goonetilleke, A.; Bandala, E.R. Treatment technologies for emerging contaminants in water: A review. *Chem. Eng. J.* **2017**, *323*, 361–380. [CrossRef]

29. Joss, A.; Siegrist, H.; Ternes, T.A. Are we about to upgrade wastewater treatment for removing organic micropollutants? *Water Sci. Technol.* **2008**, *57*, 251–255. [CrossRef]

30. Benstöm, F.; Stepkes, H.; Rolfs, T.; Montag, D.; Pinnekamp, J. Reduction of Micropollutants Using Granular Activated Carbon in an Existing Flocculation Filter on a Municipal Waste Water Treatment Plant in Düren, Germany. *8 th Micropol & Ecohazard* **2013**. Available online: https://www.researchgate.net/publication/318908086_Reduction_of_micropollutants_using_Granular_Activated_Carbon_in_an_existing_flocculation_filter_on_a_municipal_waste_water_treatment_plant_in_Duren_Germany (accessed on 8 August 2020).

31. Snyder, S.A.; Adham, S.; Redding, A.M.; Cannon, F.S.; Decarolis, J.; Oppenheimer, J.; Wert, E.C.; Yoon, Y. Role of membranes and activated carbon in the removal of endocrine disruptors and pharmaceuticals. *Desalination* **2007**, *202*, 156–181. [CrossRef]

32. Rossner, A.; Snyder, S.A.; Knappe, D.R. Removal of emerging contaminants of concern by alternative adsorbents. *Water Res.* **2009**, *43*, 3787–3796. [CrossRef]

33. Torrellas, S.Á.; Lovera, R.G.; Escalona, N.; Sepúlveda, C.; Sotelo, J.L.; García, J. Chemical-activated carbons from peach stones for the adsorption of emerging contaminants in aqueous solutions. *Chem. Eng. J.* **2015**, *279*, 788–798. [CrossRef]

34. Benstoem, F.; Pinnekamp, J. Characteristic numbers of granular activated carbon for the elimination of micropollutants from effluents of municipal wastewater treatment plants. *Water Sci. Technol.* **2017**, *76*, 279–285. [CrossRef]

35. Ahmed, M.J.; Hameed, B.H. Removal of emerging pharmaceutical contaminants by adsorption in a fixed-bed column: A review. *Ecotox. Environ. Safe* **2018**, *149*, 257–266. [CrossRef]

36. Bazan-Wozniak, A.; Pietrzak, R. Adsorption of organic and inorganic pollutants on activated bio-carbons prepared by chemical activation of residues of supercritical extraction of raw plants. *Chem. Eng. J.* **2020**, *393*, 124785. [CrossRef]

37. Hubetska, T.; Kobylinska, N.; García, J.R. Efficient adsorption of pharmaceutical drugs from aqueous solution using a mesoporous activated carbon. *Adsorption* **2020**, *26*, 251–266. [CrossRef]

38. Ravina, I.; Paz, E.; Sofer, Z.; Marm, A.; Schischa, A.; Sagi, G.; Yechialy, Z.; Lev, Y. Control of clogging in drip irrigation with stored treated municipal sewage effluent. *Agric. Water Manag.* **1997**, *33*, 127–137. [CrossRef]

39. Duran-Ros, M.; Puig-Bargués, J.; Arbat, G.; Barragán, J.; de Cartagena, F.R. Performance and backwashing efficiency of disc and screen filters in microirrigation systems. *Biosyst. Eng.* **2009**, *103*, 35–42. [CrossRef]

40. Rahman, N.; Khan, M.F. Nitrate removal using poly-o-toluidine zirconium (IV) ethylenediamine as adsorbent: Batch and fixed-bed column adsorption modelling. *J. Water Process Eng.* **2016**, *9*, 254–266. [CrossRef]

41. Kårelid, V.; Larsson, G.; Björlenius, B. Pilot-scale removal of pharmaceuticals in municipal wastewater: Comparison of granular and powdered activated carbon treatment at three wastewater treatment plants. *J. Environ. Manag.* **2017**, *193*, 491–502. [CrossRef] [PubMed]

42. De Ridder, D.J.; Villacorte, L.; Verliefde, A.R.; Verberk, J.Q.; Heijman, S.G.J.; Amy, G.L.; Van Dijk, J.C. Modeling equilibrium adsorption of organic micropollutants onto activated carbon. *Water Res.* **2010**, *44*, 3077–3086. [CrossRef]

43. Boreen, A.L.; Arnold, W.A.; McNeill, K. Photochemical fate of sulfa drugs in the aquatic environment: Sulfa drugs containing five-membered heterocyclic groups. *Environ. Sci. Technol.* **2004**, *38*, 3933–3940. [CrossRef]

44. Yang, X.; Flowers, R.C.; Weinberg, H.S.; Singer, P.C. Occurrence and removal of pharmaceuticals and personal care products (PPCPs) in an advanced wastewater reclamation plant. *Water Res.* **2011**, *45*, 5218–5228. [CrossRef]

45. Motwani, P.; Vyas, R.K.; Maheshwari, M.; Vyas, S. Removal of sulfamethoxazole from wastewater by adsorption and photolysis. *Nat. Environ. Pollut. Technol.* **2011**, *10*, 51–58.

46. Telgmann, U.; Borowska, E.; Felmeden, J.; Frechen, F.B. The locally resolved filtration process for removal of phosphorus and micropollutants with GAC. *J. Water Process Eng.* **2020**, *35*, 101236. [CrossRef]

47. Sotelo, J.L.; Ovejero, G.; Rodríguez, A.; Álvarez, S.; Galán, J.; García, J. Competitive adsorption studies of caffeine and diclofenac aqueous solutions by activated carbon. *Chem. Eng. J.* **2014**, *240*, 443–453. [CrossRef]

48. Ang, T.N.; Young, B.R.; Taylor, M.; Burrell, R.; Aroua, M.K.; Baroutian, S. Breakthrough analysis of continuous fixed-bed adsorption of sevoflurane using activated carbons. *Chemosphere* **2020**, *239*, 124839. [CrossRef] [PubMed]

49. Decree, S.R. 1620/2007, 7 December 2007. Spanish Regulations for Water Reuse. Real Decreto, 1620. Available online: https://www.boe.es/buscar/doc.php?id=BOE-A-2007-21092 (accessed on 28 June 2020).

50. Chae, S.H.; Kim, S.S.; Jeong, W.; Park, N.S. Evaluation of physical properties and adsorption capacity of regenerated granular activated carbons (GACs). *Korean J. Chem. Eng.* **2013**, *30*, 891–897. [CrossRef]

51. Government of Spain website. Ministry of Environment and Rural and Marine Environment. National Water Quality Plan: Environmental Sustainability Report. 2010. Available online: https://www.miteco.gob.es/images/es/isa_pnra_231210_tcm30-183615.pdf (accessed on 31 July 2020).

52. Jiménez, B.; Asano, T. (Eds.) *Water Reuse: An International Survey of Current Practice, Issues and Needs*; IWA: London, UK, 2008.

© 2020 by the authors. Licensee MDPI, Basel, Switzerland. This article is an open access article distributed under the terms and conditions of the Creative Commons Attribution (CC BY) license (http://creativecommons.org/licenses/by/4.0/).

MDPI

St. Alban-Anlage 66

4052 Basel

Switzerland

Tel. +41 61 683 77 34

Fax +41 61 302 89 18

www.mdpi.com

Water Editorial Office

E-mail: water@mdpi.com

www.mdpi.com/journal/water

www.ingramcontent.com/pod-product-compliance
Lightning Source LLC
LaVergne TN
LVHW070610170726
843515LV00004B/38